THWART CLIMATE CHANGE NOW

Reducing Embodied Carbon Brick by Brick

By Bill Caplan

ENVIRONMENTAL LAW INSTITUTE
Washington, D.C.

Dedicated to
Common Sense

Table of Contents

Prologue

"The most alarming of man's assaults upon the environment is the contamination of the air, earth, rivers and sea with dangerous and even lethal materials."[1] Rachel Carson's words in *Silent Spring* referred to chemicals, the massive use of pesticides. The year was 1962. *Silent Spring* provided a catalyst for the modern environmental movement more than a half-century ago spawning grassroots activities across the country. A new consciousness awakened that led to the first Earth Day on April 22, 1970. Twenty million Americans participated in calling for a green environment, and to this day, Earth Day events are celebrated annually around the world. The U.S. Environmental Protection Agency was also founded that year, "born in the wake of elevated concern about environmental pollution."[2] While 2020 marked the 50th anniversary of both, we still face the reality of our current assault on the environment—greenhouse gas (GHG) emissions—as we approach Earth Day 2022.

During those 50 years, we learned that gases emitted while burning carbon-based fuels are just as dangerous to our survival as the indiscriminate use of pesticides revealed in *Silent Spring*. Not only emissions from powering transportation, lighting, heating, cooling, and industrial production, but also those associated with fabricating buildings and infrastructure—from mining and processing raw materials to final construction. Unfortunately, as in 1962, we have been assaulting the environment through complacency. Yet since that time, our drive and technological prowess carried people to the moon and back and remotely directed vehicles across the surface of Mars. Despite our engineering capability and knowledge of sustainable design we ignored the detrimental effects of consumption and pollution, and continue to construct our built environment with such disregard. We continue to consume our planet's resources at an expanding rate, producing waste and pollution in ever-increasing quantities. This is especially egregious considering the commercial availability of clean energy from solar, wind, geothermal, and hydrothermal sources, and because architects, planners, and policymakers have been educated in environmental design strategies. Science and engineering have brought numerous energy-efficient building products to fruition as

---

1. Rachel Carson, Silent Spring 6 (1962) [hereinafter Carson].
2. EPA History, https://www.epa.gov/history (Dec. 18, 2018).

well as a fivefold increase in insulating capabilities. Though aware of global warming and the impact of GHGs, society continues to ignore the ecological impact of its creative output—of what we design, fabricate, and build. What is wrong with this picture? In 1962, Carson noted the irony that "man might determine his own future by something so seemingly trivial as the choice of an insect spray."[3] Today, ironically, while facing the current crises of global warming and pollution, we draw comfort from green and sustainable design labels, rather than their veracity and efficacy which are the very key to humanity's future. Our love affair with the notion of what we call sustainable and green is palliative. Their mere mention sparks approvals rarely questioned, often feckless like the *Emperor's New Clothes* but regrettably less transparent. It is time to wake up. Too often, the buildings and products we design and construct fail to constitute "sustainable" design regardless of their certification, recognition, or awards received. Too often, their materials and fabrication overuse resources, produce excessive waste, generate pollution, or function inefficiently. This need not be so.

Growth of the built environment presents a significant problem—from new buildings to renovations to retrofits, from temporary structures to infrastructure. After all, what we build today will stand for 50 to 100 years. With 8 billion square feet (7.7 billion m^2)[4] of new construction projected each year through 2030, more than 11 billion metric tons of carbon dioxide (CO_2)[5] will be released to the atmosphere from building operations and new construction—annually. This is nearly 40% of all energy-related carbon emissions worldwide. We have the opportunity to reduce such emissions significantly through serious sustainable design. Given the prevailing knowledge of sustainable and green design techniques, our technical capability, and the media's broad coverage of architectural achievements, one might assume that meaningful emissions reductions have been achieved. Although the proliferation of awards, certifications, fawning reviews, and art installations convey that impression—it is simply not so. Despite substantial gains achieved through the use of light-emitting diodes (LED), energy-efficient appliances, and renewable energy, gains derived from the science of building are severely lacking, with addressing "embodied carbon" emissions among the most overlooked. This failure to take the fundamentals of sustainable design

---

3. CARSON, *supra* note 1, at 8.
4. INTERNATIONAL ENERGY AGENCY (IEA), PERSPECTIVES FOR THE CLEAN ENERGY TRANSITION: THE CRITICAL ROLE OF BUILDINGS (Apr. 2019).
5. IEA & UNITED NATIONS ENVIRONMENT PROGRAMME, 2018 GLOBAL STATUS REPORT: TOWARDS A ZERO-EMISSION, EFFICIENT, AND RESILIENT BUILDINGS AND CONSTRUCTION SECTOR (2018). According to the report, "total buildings-related CO_2 emissions amounted to more than 11 GtCO_2 in 2017." 1 Gt = 1 billion metric tons.

seriously is long overdue for correction. Whether or not Nero fiddled while Rome burned is academic; our present fiddling with future solutions while temperatures rise is real. *Thwart Climate Change Now: Reducing Embodied Carbon Brick by Brick* seeks to shed light on the built environment's unseen emissions and lay a pathway to their reduction "now," before a precarious concentration of atmospheric carbon has been set in concrete. In 1910, the atmospheric concentration of CO_2 reached 300 parts per million (ppm) for the first time in more than 300,000 years. It averaged only 317 ppm in 1960 while Rachel Carson was penning *Silent Spring*—increasing merely 17 ppm over 50 years. The next 50 were less gentle. By 2010, the increase in CO_2 concentration was four times that, and on its way toward 417 ppm in 2021.[6] Keep that in mind as you read this book.

Bill Caplan
September 2, 2021

---

6.   Earth System Research Laboratories' Global Monitoring Laboratory of the National Oceanic & Atmospheric Admin., Ed Dlugokencky & Pieter Tans, *NOAA/ESRL*, www.esrl.noaa.gov/gmd/ccgg/trends. Data for 1910 and 1960 is: *Ice Core Data Adjusted for Global Mean, in* NASA GODDARD INSTITUTE FOR SPACE STUDIES: FORCINGS IN GISS CLIMATE MODE, WELL-MIXED GREENHOUSE GASES, HISTORICAL DATA (2014).

Introduction

Metaphorically, the path to sustainability was laid "brick by brick."
Buildings emerge "brick by brick."
So do carbon emissions.

Thwart Climate Change Now: Reducing Embodied Carbon Brick by Brick addresses an imperative—to slow the pace of climate change within the upcoming decade; now rather than in the distant future. This can be achieved by significantly reducing the carbon footprint of the "physical" environment we construct and renovate. When future energy supplies worldwide approach *"carbon-free,"* atmospheric carbon levels should stabilize—*but what do we do until we get there?* With temperatures continuing to rise, people around the world have taken note. The handwriting is on the wall; human activity has impacted the earth's environment; our climate is already changing. Accordingly, 196 nations signed the United Nations Framework Convention on Climate Change (UNFCCC) known as the Paris Agreement:

> [T]o strengthen the global response to the threat of climate change [by] holding the global average temperature increase to well below 2°C above pre-industrial levels while pursuing a lesser increase of only 1.5°C.[1]

How? By reaching the "global peaking of greenhouse gas emissions as soon as possible" and undertaking "rapid reductions thereafter"; by achieving "a balance between anthropogenic emissions by sources and removals by sinks of greenhouse gases in the second half of this century"; all within "the context of sustainable development and efforts to eradicate poverty."[2] As such, policymakers, scientists, and engineers around the globe are pursuing the means to limit global warming to less than 2 degrees Celsius (°C) (3.6 Fahrenheit (°F)) while targeting a temperature rise of no more than 1.5°C (2.7°F). Reducing the emissions from electricity generation and the consumption of energy by transportation, agriculture, construction, and building operations has become paramount. Power sources produce more than two-thirds of all global emissions, 40% of those owing to the energy consumed and the byproducts released by *manufacturing building materials, constructing buildings, and their post-occupancy operations.* In short, our built environment represents the largest contributor to greenhouse gas (GHG) emissions of any sector other than the production of energy itself. Residential buildings—providing shelter—comprises its largest share.[3]

Although society has awakened to the need to address our future through green and sustainable actions, too often the *future-addressing activities* themselves provide a false sense of progress, and with some, more lip-service than follow-through. Beware of manufacturers' public relations and Internet misinformation. Beware of platitudes, awards, and certifications. Beware of the cover of labels such as "sustainable" and "green." Beware of fictitious claims. Many wishful solutions ignore their own inherent environmental cost. Even seemingly emissions-free energy, such as from solar panels, emit greenhouse gases (GHGs) during their manufacture. The same applies to energy-conserving solutions such as triple-pane windows. The elephant in the room is invisible, an *embodiment* of carbon emissions—embodied carbon; the carbon footprint of everything we construct and manufacture. Not a physicality itself, embodied carbon is the realization of a building's carbon footprint from cradle through construction to completion—of GHGs *already emitted, already contributing* to global warming. It also represents the recognition of emissions to come from a building's maintenance and eventual demise. Embodied emissions must be the first tally of sustainability when weighing success or failure—they already abide in our atmosphere, heretofore absorbed. This holds true for every new structure built and every old structure renovated or replaced. By the time a building is ready for occupancy, most of its embodied emissions have been released to the environment. Because this is *fait accompli* on a building's first day of operation without recourse, and it increases the underlying rate of global warming, minimizing embodied carbon in buildings still on the drawing board or already in construction must take priority. Yes, the ultimate solution of zero-carbon energy is on the horizon, achieving that goal on a global scale remains several decades away. Because of that reality, lessoning the buildup of "embodied" carbon at this moment is essential. We must curtail its escalation until the gradual implementation of carbon-free energy gains meaningful traction on a worldwide scale. Embodied carbon is the poison in our dwellings, and in everything else we construct or manufacture.

This problem extends far beyond the growth and renewal of our existing urban and suburban areas; it prevails magnified in developing nations worldwide and in refugee settlements as well. The path from substandard

---

3.   Data and computations from United Nations Environment Programme, Emissions Gap Report 2019 (Nov 2019).

to permanent housing progresses from wood, corrugated metal, and tent structures to more impactful concrete block and brick. The initial increase in embodied emissions to upgrade dwellings for more than a billion poorly housed people will be substantial. The movement of populations from mountainous regions, river valleys, shorefronts, and farmlands to ever-growing urban areas will in turn require vertical housing. Unfortunately, as embodied emissions are related to a building's floor area, they increase exponentially with each added floor of a high-rise residential building. Improved cooking facilities and sanitation will add more, especially for the billion people currently without access to electricity. Little of this has been factored in the mitigation solutions at hand, nor in the modeling of future cities. While we envision futuristic urbanization through a low-carbon lens we cannot ignore its coincident explosion in energy consumption and the emissions embodied in construction. They will continue to grow. With scientists warning we have little more than a decade to decrease green-house emissions, we must deal with such curtailment today. As the days of reckoning become more apparent, parsing the efficacy of our near-term actions remains elusive. Addressing tough choices with platitudes rather than substantive action has become the norm, even when confronted with an existential threat.

Creating shelter for survival dates back to early humans and environmental design has been with us in one form or another for several millennia, but the need to address *humanity's impact* on the environment has surfaced only recently. While sustainable and green thinking has been gaining traction for 50 years, the technical prowess to effectively implement their precepts is relatively new. Unfortunately, the commercial exploitation of the terms *sustainability* and *green* has blossomed to obscure their meaningful application. The paths to tackle carbon emissions rely on technology, policy, and politics requiring a long timeline to navigate, with little opportunity to short circuit. Nevertheless, the built environment emerges one building at a time, both large and small. Although total emission reductions depend on the future availability of low-carbon power, the built environment's embodied carbon can be reduced fairly quickly—brick by brick: by the developers, architects, engineers, and designers that dictate its structure, materials, and environmental interface. Attention to embodied carbon in the *design* of our buildings and infrastructure, as well as attention to the impact of embodied carbon in "green" policies and incentives on a local level, can reduce carbon emissions within a decade.

Part One, *Sustainability*, traces the emergence of sustainable design as we have come to know it—its roots, how one perceives it, its efficacy, its role as placebo, and our current quest for zero-carbon energy. Commencing with a reality check, Part One provides the background material and perspectives that illuminate our current situation, elaborating on the development of environmental design, sustainability and green, and the nuances of carbon emissions. It closes with a call for action that addresses the urgency of thwarting climate change now, without waiting for the long-term solutions. The first chapter, Buzzwords & Muddle, reveals sustainable design to be more elusive than one would think. Cherry-picked facts proffer misleading illusions and published assertions are more muddy than clear. Chapter Two, From Instinct to Science; Philosophy to Practice, provides historical insight into the *environment* functioning as a vector of design and the quest for energy conservation over the last three decades. Chapter Three, Parsing Carbon, Sustainability, and Green, elucidates the subtleties of carbon emissions and the concepts of "sustainability" and "green" to add clarity to the current issues at hand. Reality and the Call for Action—the final chapter—explores the issues and dilemmas that must be addressed to retard emissions-induced global warming in real time, while we wait for the future solve-all, zero-carbon energy worldwide.

Part Two, *Thwarting Climate Change*, lays a path forward to reduce the matter-of-fact acceptance of embodied carbon. Clearly stating what we are up against, as well as the role of design decisions, Part Two addresses the emissions impact of design as well as the tools, policies, and legislation needed to reduce net carbon emissions within a decade. The first two chapters, What We Are Up Against; What Must We Do (Chapter Five) and The Role of Design (Chapter Six) target near-term emissions goals by the means of material choice and design methodology, approaches available to every developer and client, and every architect, engineer, and designer. Chapter Seven, Confronting Embodied Carbon Now focuses on the power of policy to reduce embodied carbon emissions throughout the upcoming decade. The discourse addresses the unintended consequences from policies that ignore embodied carbon emissions that fail to consider embodied carbon while calculating their presumed benefits. The final chapter, Taking the Roads Less Traveled, refers to Rachel Carson's rendering of Robert Frost:

> The road we have long been traveling is deceptively easy, a smooth superhighway on which we progress with great speed, but at the end lies disaster. The other fork of the road—the one "less traveled by"—offers our last, our only chance to reach a destination that assures the preservation of the earth.

The choice, after all, is ours to make.[4]

Taking the Roads Less Traveled provides policy and legislative actions that are critically needed to reduce embodied carbon from now through 2030-2035.

Many terms currently in use describe or indicate the presence or emission of GHGs such as the abbreviations, symbols, and acronyms GHG, CO_2, CO_2eq, "carbon dioxide," "embodied carbon," "carbon footprint," or simply "carbon." They all refer to gases that capture radiant thermal energy and re-radiate a part that warms the earth's atmosphere and surface. Carbon dioxide gas (CO_2), is by far the most detrimental. Methane, although more potent, is emitted in lower volume and primarily associated with agriculture, animal husbandry, and landfills. Nevertheless, the large leakage of methane gas that occurs during natural gas extraction is receiving more scrutiny. Technical studies and scientific papers typically include the impact of methane and other GHGs by rating their "equivalent" global warming potential (GWP) to that of CO_2. The symbol "CO_2eq" serves to group those GHG emissions with CO_2 in order to indicate their emissions collectively. All energy-related GHG emissions referenced in this text, including CO_2eq, are collectively referred to as "carbon dioxide," "embodied carbon," "carbon footprint," "greenhouse gases," or simply "emissions" or "carbon" in order to provide a more legible discussion for a broad readership.

The emission statistics presented in this book have been sourced or computed from a variety of public documents issued by government agencies, international commissions, institutions, and academic studies. As such, small variances may appear due to the differential between CO_2 and CO_2eq, differing term definitions, computation methodologies, or mathematical rounding, none of which alter the conclusions, recommendations, or relevance of this discourse.

In its broadest definition, "embodied carbon" refers to all carbon gases emitted during processing, manufacturing, and fabricating everything we produce. This includes raw material extraction, transportation, installation, maintenance, and a product's eventual discard. Some emissions result from the energy consumed when using carbon-based fuels and others derive from chemistry changes during material processing. Expressed in the mass of emissions released in kilograms or pounds, "embodied carbon" really means "embodied emissions." Internationally, GHGs are expressed in tonnes, a term also referred to as metric tons. One tonne (metric ton) equates to 1,000

---

4.   RACHEL CARSON, SILENT SPRING 277 (1962).

kilograms, 2,200 pounds, or 2.2 tons in the United States. A metric ton (tonne) is 10% more than an imperial ton. A gigatonne (Gt) constitutes one billion tonnes or one billion metric tons.

Thwart Climate Change Now: Reducing Embodied Carbon Brick by Brick emphasizes immediacy, acting "now," achieving significant reductions in carbon emissions within 10 to 15 years—from 2020 through 2035. Scientific consensus warns that we must throttle emissions growth within this period before their immediate warming impact will make capping even a 3°C temperature rise unlikely. "Now," meaning *as quickly as possible*, addresses the upfront emissions inherent in producing the materials we select and caused by our design and construction methodologies—those emitted long before society benefits from operational efficiencies or achieves the long-term goal of a carbon-free energy supply. *Thwart Climate Change Now* explains why some green initiatives are too broadly incentivized and may exacerbate current conditions; and why some energy-efficient upgrades, residential solar panel installations, and net-zero buildings might contribute to global warming within their first decade of operation, and therefore should wait. Most importantly, *Thwart Climate Change Now* is a wakeup call for immediate actions that can be initiated with the design of every new building, renovation, or retrofit.

Part One
Sustainability and Green

Chapter 1

Buzzwords & Muddle

A tree grows in Brooklyn.
Cut it down.

After 35 years in the high-tech industry I enrolled in architecture school. The year was 2006. High-tech had taken us to the moon and back and had recently landed a probe on Saturn's moon Titan. Science and engineering commercialized solar panels and energy-efficient building materials. And architects had been addressing issues of sustainable design for over 15 years. We had the knowledge, we had the tools, we had the insight. The "Green movement" was well entrenched. People cared. Yet despite centuries of building experience, decades of environmental enlightenment, and spaceflight-worthy advances in science and engineering, urban design seemed stuck in the mud—not particularly friendly to people, not especially concerned with the environment. This was notably so in the United States. Unobservant or unaware, the media portrayed a different picture, one of dynamic progress in architectural design. Awards proliferated at an increasing rate. Yet, greenhouse gas (GHG) emissions continued to mushroom; and it was common knowledge that buildings were responsible for a major share. What was wrong with this picture, this alternate world? How much did the public know? How much did the design profession itself understand? So began my quest for an education in architecture—to discover "why" this was happening and to see what could be done about it.

Reality was revealed rather quickly in the second semester, interestingly enough from a border of trees growing in Brooklyn. Similar to the symbolism in the namesake book,[1] these trees grew amidst concrete and asphalt, giving life to an otherwise congested area, uplift and oxygen to a chaotic urban environment, and momentary peace to people from all walks of life. The trees bordered a small triangular park with an undulating green landscape

---

1. Betty Smith, A Tree Grows in Brooklyn (1943).

3

bounded by bustling traffic on all sides. A third of an acre gem, this patch of green was reclaimed from the removal of an unsafe building 20 years prior. Visit the park for inspiration for a theme; sketch your idea; present it at our next studio pin-up—these were the instructions. Inspired by this serene retreat of nature, I sketched a swirl above the ground, the green canopy bordering the triangular lot, partially open on the front exposure, envisioning a community performance space integrated with the mature trees—a design that preserved the "green." The critique came quickly: "No trees—cut them down." Ripped from the wall, crumbled in a ball, the sketch was tossed to my drafting table for emphasis. "No trees." This was about building. At the juried mid-term, a professor of landscape architecture concurred: bulldoze the lot. Though it is unquestionably advantageous to build on a cleared flat lot, their professorial certitude was never explained. Perhaps a requisite of "green" burdens a project's focus, distracts from a client's objective. Air conditioning can replace a tree's cooling shade, sunscreens can break the sun's glare—and—there are other ways to become "green." All of this is true, but not particularly sustainable. If the fiat had been "no concrete" rather than "trees," we would be better off today. Some things carry more weight than others.

Throughout those years, sustainability and green were nothing more than buzzwords when it came to building design. Buzzwords vaguely promoting recycling and reuse, solar panels and geothermal, energy star appliances or pouring concrete onsite rather than trucking blocks and bricks from afar. Although such practices might be laudable for some design parti, they do not constitute *sustainable building design*, the key to a low carbon built environment, a key to reducing the rate of global warming. Throughout those years, though a component of discussion in every studio, sustainable and green were gift wrapping for an appearance of green, whose impact lacked relevance—a cover for not addressing cause and effect, for not examining the magnitude of the redress for things that matter, for not analyzing the facts. Facts matter; buzzwords obscure.

We have reached the 2020s and not much has changed. Buildings are taller, appliances are more efficient, and we burn less coal, but architecture's failure to synergize with the environment—to take sustainable design seriously—has not changed.

The Muddle

Headline news was not particularly good in early January 2019 when carbon emissions were a significant theme. The Rhodium Group had released its *Preliminary U.S. Emissions Estimates for 2018* and the numbers were not pleasing.

U.S. Carbon Emissions Surged in 2018
Even as Coal Plants Closed
Brad Plumer, New York Times

U.S. Greenhouse Gas Emissions Spiked in 2018
— and It Couldn't Happen at a Worse Time
Chris Mooney and Brady Dennis, Washington Post

Emissions of Carbon Climb 3.4%
Kris Maher, Wall Street Journal

The *New York Times* noted that "manufacturing boomed" and "emissions from factories, planes and trucks soared," particularly from the industrial sectors such as "steel, cement, chemicals, and refineries," and a "relatively cold winter led to a spike in the use of oil and gas for heating." The month before, *U.S. News & World Report*[2] noted that the unusually warm summer weather, requiring cooling, also added to the mix. The *Wall Street Journal* stated that the "U.S. has been among the world's few success stories, dramatically reducing emissions" by moving away from coal-fired power, but the current level of emissions have "reversed that trend" with the largest increase since 2010 coming after three straight years of declines.[3] The *Washington Post* called this a "jarring increase that comes as scientists say the world needs to be aggressively cutting its emissions to avoid the most devastating effects of climate change."

Nonetheless, they all noted that even with this increase, U.S. carbon emissions in 2018 were lower than emissions in 2005. According to the Rhodium Group, 11.2% less. Annual increases are always unwelcome news, but overall emissions 11% below the prior 12-year low—isn't that pretty good? *The facts matter—but what were they?* What was behind those numbers, what do we need to focus on? Headlines are sound bites lacking explanatory information. Unfortunately, the articles themselves were often confusing—often misleading—often missing the point.

A month earlier, an announcement directed to architects, designers, and builders was posted to the web portal of the American Institute of Architects (AIA) Committee on the Environment, automatically relayed to its members

2. Alan Neuhauser, *Global Emissions Climb to Record Highs, Reversing Three Years of Declines*, U.S. News & World Report, Dec. 5, 2018.
3. Kris Maher, *Emissions of Carbon Climb 3.4%*, Wall St. J., Jan. 9, 2019, at A3. The article was sourced from a Jan. 8, 2019, post—https://rhg.com/research/preliminary-us-emissions-estimates-for-2018/.

by email. Based on a prior report of the same information, this one provided some upbeat news for architects.

It began:

Dec. 13, 2018:

Happy Holidays!
THIS IS BIG

Amid all the sobering stories and projections about climate change in the news lately, we have some upbeat news to share. Our hard work is having a BIG impact.

Today, U.S. building sector CO_2 emissions are 20.2% below 2005 levels.

According to data from the U.S. Energy Information Administration, energy efficiency and power sector decarbonization have reduced U.S. building sector CO_2 emissions by 20.2% below 2005 levels, despite adding approximately 30 billion square feet to our building stock during the last 12 years.

And, global building sector CO_2 emissions appear to have leveled off in the past few years.

That's the good news.

Of course, that's only the beginning. There is still much, much more to do. . . .[4]

Imagine, although we constructed an additional 30 billion square feet of new buildings in the United States over the prior 12 years, overall building sector carbon emissions now were 20% less than in 2005. We must be doing something right. Astonished by this "upbeat" news, I paused to rethink this book. But when I had time to digest the email's remaining pages, the message went from upbeat to mixed, from confusing to dour. Researching the source material opened a Pandora's box. A plethora of mixed data, definitions, cross-references, and interpretations—a literal can of worms. No wonder the reality is obscure, leaving journalists to choose from conflicting presentations and apparent contradictions. No wonder conclusions require numerous and confusing qualification. On a quick read, you might have found the 20.2% reduction cited above confusing; the Rhodium Group reported 11.2%. But one refers to the *building sector alone*, while the other to all energy-related carbon dioxide (CO_2) emissions. It can be hard to keep track of the facts without redoing the math. After a while, they get lost in the muddle.

---

4.   Posted to the AIA Committee on the Environment Digest on Dec. 13, 2018, by Edward Mazria FAIA sourced from the Architecture 2030 E-NEWS (Emphasis in original.).

From the perspective of energy-based carbon emissions, most reporting identifies the transportation sector as the largest contributor. While transportation does loom large, carbon emissions related to the built environment are significantly more voluminous, a fact not easily deciphered. Transportation is one of the major contributors, responsible for one-quarter of the global emissions[5] and a well-publicized target for emissions reduction. Such awareness has spawned the movement toward alternative fuels, electric vehicles and smaller cars. On the other hand, a vague understanding of the built environment's role has engendered a false sense of progress, promulgated by our romance with all things "sustainable" and "green" without a means to evaluate their worth. Those who try face a challenging task. Evaluating emissions attributable to the built environment entails at least two economic sectors, buildings and construction. Some studies include one or the other or both, others track only building operations. Many ignore construction vehicle emissions and emissions from transporting materials. As diverse models define each reporting agency's presentation, their disparate use classifications, footnotes, and fine print muddle the very picture they purport to present. We understand that green and sustainable buildings are important, as are green vehicles and green renewable power—but not necessarily how they relate to each other, or what they entail. *Sustainable habits, sustainable design, eliminating fossil fuels*: all three make a difference. But at this stage of our battle to reduce global warming, which need more attention? What is the reality? The facts matter.

Two numbers compiled from the source material stand out like red warning lights, derived from statistics provided by the International Energy Agency and the United Nations Environment Programme. As numbers they appear innocuous, but as a percentage of total energy-related CO_2 emissions they are stunning. Both relate to our built environment: 40 and 7.

Approximately 40 percent of all energy-related carbon emissions result from the construction and operation of buildings.[6]

The manufacturing of a single material alone accounted for 7 percent of all energy-related carbon emissions worldwide—cement.[7]

---

5. https://www.iea.org/tcep/transport/.
6. IEA & United Nations Environment Programme, 2018 Global Status Report: Towards a Zero-Emission, Efficient, and Resilient Buildings and Construction Sector (2018) [hereinafter 2018 Global Status Report].
7. "The cement industry currently represents about 7% of the carbon dioxide (CO_2) emissions globally": *Technology Roadmap Low-Carbon Transition in the Cement Industry*, a collaborative effort between the IEA and the Cement Sustainability Initiative CSI of the World Business Council on Sustainable Development in 2018.

Let that sink in—it's our shelter, the built environment; emissions from building construction and building operations. There is a panoply of emission sources to decode within this 40% share, but cement tops the list—a singular low hanging fruit. The most widely used building material in the world is a huge polluter, a major contributor to global warming. Yet this constitutes merely one constituent in a broad category known as "embodied carbon"—a topic of considerable concern that may be noted, though in practice mostly ignored. The implementation of "sustainability" and "green" require our scrutiny. There lies the means to achieve synergies among people, buildings, and the environment—but only if we understand what they are based upon, what they mean and what actions matter most now.

Eliminating Fossil Fuels Versus Sustainable Design

When zero-carbon sources generate most of the world's energy, global warming from carbon emissions will no longer be an issue. Building construction and operation, transportation, and industrial production will emit minimal carbon, only the emissions inherent from a product's chemistry will remain. If you are confident that fossil-fuel energy will be eliminated in time to stabilize global warming, read no further—just wait it out. If you are unsure, consider the following: progress has been made by substituting low-carbon fuels and zero-carbon sources for coal and oil, but there is a long road ahead. Even with the promise of electric vehicles, emissions still default to the source of their recharge, electricity generation. Wind, solar, and battery technology are making significant gains in addition to the old standards of hydroelectric and nuclear; geothermal and hydrothermal energy offer potential as well. Nonetheless, carbon-based fuels, which include biomass, are estimated to generate nearly 91% of the world's current energy consumption. Carbon-free energy amounts to only 9% worldwide.[8] We have a long way to go.

In the coming decades, global urbanization will exacerbate the need for clean carbon-free energy. As of 2017, nearly 1 billion people lived without electricity and 2.7 billion people lacked clean cooking facilities.[9] As our built environment expands to alleviate those conditions the demand for energy will increase. We will likely consume more fossil fuel than can be replaced by carbon-free sources for decades to come. Eliminating fossil fuels is an imperative—but *what we do until we get there* is equally important, or perhaps more so. Unless we change the way we design and build now, too much of our built environment will have been built before the world achieves substan-

---

8. IEA, Renewables Information: Overview (2019).
9. 2018 Global Status Report, *supra* note 6.

tial zero-carbon energy generation. This means that too many gigatons of emissions will have been released to the atmosphere, or locked-in for future releases, to hold temperature increases to anywhere near 2 degrees Celsius. Fortunately, science and technology have stepped up to the plate, creating energy-efficient vehicles, appliances, lighting, and building materials. Fortunately, public awareness of "green" living is growing, encouraging us "reduce, reuse, and recycle." Nonetheless, while carbon fuels still reign and embodied carbon emissions are mostly overlooked, the unattended pathology resides in how we design and construct our built environment. At best, this is misunderstood; at worst, mostly ignored. This book focuses on architecture's impact on global warming because the veracity of "sustainable design" needs much more of our attention. The move toward low-carbon and no-carbon energy—though slow and a long haul—is already mainstream. Verifiable sustainability of the physical environment, however, is more elusive than one would think, and too often mere illusion. Given media accolades for new architecture, trade group awards, environmental certifications, and our passion for "green," this might seem puzzling. So let's play it safe: "When a tree grows in Brooklyn," let it prosper.

Chapter 2

From Instinct to Science; Philosophy to Practice

Rolling Back the Clock

Reading a book about the history of architecture or a manifesto on architectural design, one likely encounters a "story of origin" that speculates how early humans created their first dwelling. Dating back more than 2,000 years, by the late 1750s these prototypical dwellings became known as the "primitive hut." We credit Vitruvius, writing in the first century BC, for the earliest surviving treatise on architecture. His manifesto *De architectura (The Ten Books of Architecture)* included such a chapter, "The Origin of the Dwelling House." I too refer to those dwellings, not to set the stage for an architectural theory, but to spotlight humanity's primal quest for safe and healthful shelter, in harmony with the natural environment. A natural environment comprised of air, water, soil, and vegetation that knits life-giving sustenance sufficient for all. When the environs stopped giving, the inhabitants simply moved on.

Whether such stories commenced with assembling protective walls of stone or mud, sheltering in a dugout, or a construction from branches and vegetation the motivation was always the same, survival: personal survival or family survival. For the first time humans rearranged their locale. Instinct and observations of nature were at work; so too were observations of other animate species and how they sheltered. This was not a philosophical quest, but an innate drive to exist; to prosper in a natural environment despite the inherent ecological challenge. That same thrust drives "sustainable design" today with one big difference. Today, our human-built environment is more than a local rearrangement. It has altered nature's equilibrium, eroded its life-giving bounty. We cannot simply move on.

The art and science of building developed slowly over the 2,000 years from the 1st century BC to the early 20th century. Not to diminish the significance of Vitruvius' treatise on architecture, it is notable that three centuries earlier, Xenophon attributed climate conscious design suggestions to Socrates. It was a matter of observation, a matter of logic.

Is it pleasant, then, to have one cool in summer and warm in winter? . . . .

Accordingly, in houses looking toward the south, does the sun shine into the inner rooms in winter, while in the summer when it travels over us and the roofs, it provides shade? If then it is noble that these things come to be thus, one should build a house higher on the southern side, so that the winter sun won't be shut out, and closer to the ground on the northern side, so that the cold winds won't burst in.[1]

This forerunner to environmental design made sense.

In *De architectura*, Vitruvius recommended orienting dwellings harmonized to the light and heat of the diurnal and seasonal solar cycles, such that bedrooms receive eastern light, baths receive western light for winter comfort, and picture galleries receive northern light. He applied similar principles to town planning, considering wind direction and climatic conditions, recommending the foresight to align streets and alleys to exclude the wind.[2] Employing the insulating properties of the earth and careful selection and preparation of building materials also dates back to ancient times. This made sense then as it does now. Throughout the history of constructing shelter, survival came first, then comfort, and eventually ornamentation and style. Whether by instinct or thoughtful observation, synergistic design has been around for a long time.

The science of building design has evolved from primitive savvy, to conventional wisdom, to sound principles. But a conscious practice of environmental design did not gain ground until the last half of the 20th century, following an earlier body of work from the 1920s through the 1940s. Some prominent architects, notably Frank Lloyd Wright, Charles-Édouard Jeanneret-Gris (known as Le Corbusier), Marcel Breuer, and Richard Neutra experimented with site and room orientation in conjunction with the sun's path, natural light, and the geometry of roofs and facade. Corbusier developed a particular interest in geometry for beneficial shading and the use

---

1.   XENOPHON'S MEMORABILIA, bk. III, ch. 8 (Amy L. Bonnette trans., Cornell University Press 2001) [hereinafter XENOPHON'S MEMORABILIA].
2.   VITRUVIUS, THE TEN BOOKS OF ARCHITECTURE 15 & 24 (Morris Hicky Morgan trans., Dover Publications Inc. 1960) (unaltered and unabridged republication of the first edition published by the Harvard Press 1914) [hereinafter VITRUVIUS, THE TEN BOOKS OF ARCHITECTURE].

of sunscreens, *brises-soleil*. Many such designs reflected on regional customs and historical attempts to contend with the sun's heat and glare, wind, rain, and the need for ventilation—a modern learning curve that enhanced this body of knowledge. Other architects, less known today, also left substantial legacies through their experimentation and writing which tethered architecture and planning to the influences of climate and the environment. We see this in Jeffrey Ellis Aronin's *Climate & Architecture* in 1953, the Olgyay brothers' *Solar Control and Shading Devices* in 1957, and Victor Olgyay's seminal work *Design With Climate* in 1963. These works contained detailed drawings, graphs, and photographs to lay out principles and methodologies that orchestrated building design in harmony with the forces of nature: solar energy, wind, precipitation, and natural ventilation. Presciently, Olgyay emphasized the importance of doing so in the *first* phases of design, not as an afterthought or attempt at refinement. "Architectural expression" would follow.[3] His design approach incorporated orientation and materials at the regional and local level, considering meteorology and solar movement as well as the natural environment. Olgyay termed this "bioclimatic architecture." The broad fundamentals of bioclimatic and environmental design stemmed in part from those age-old techniques that subordinated the design process so as to benefit from the properties of nature. But even far into the 20th century, society saw no pressing need to conserve energy, employ renewable resources, or contain pollution by means of the built environment's design.

Not until the 1950s and 1960s, with a number of modernists incorporating bioclimatic principles in their strategies, and with environmental pollution on the rise, was the groundwork for environmentally conscious architecture[4] set. Noxious vehicle fumes, industrial emissions and a major oil spill were polluting the air, water, and shorelines. The 1962 release of Rachel Carson's *Silent Spring* exposed the unfettered use of DDT and toxic chemicals on farmlands, gardens, forests, and homes, noting that over 200 such chemicals were created since the mid-1940s. Widely read, *Silent Spring* was a wakeup call to our collective role in the poisoning, lullabied by misinformation and the sway of inaction.[5] Though Carson's alarm piloted the movement toward "green," the laser focus was on chemical toxicity. Our broader understanding of sustainability and sustainable design remained decades away.

The notion of being *green* or of *green design* embedded within popular culture long before the concept of "sustainability" was defined. Use of the

---

3. VICTOR OLGYAY, DESIGN WITH CLIMATE: BIOCLIMATIC APPROACH TO ARCHITECTURAL REGIONALISM (Princeton Univ. Press 2015) (new and expanded edition).
4. Shady Attia, professor of sustainable architecture and building technology at the Université de Liège.
5. RACHEL CARSON, SILENT SPRING 7 (1962).

term "green," a color associated with nature and natural living, increased in concert with the growth of the environmental movement during the 1960s. "Green" quickly became a popular moniker for anything associated with nature or naturalness, healthy or clean living, as with the terms "organic" or "natural" used in conjunction with cotton, fertilizer, dietary supplements or food, to name a few. The "green" label cast certain unhealthy chemical compositions as less unhealthy—such as paints, adhesives, wood conditioners, and sealants that do not release volatile organic compounds. Products that "do no harm" or conservation practices such as recycling, re-use, and rainwater capture purport to be green. Solar or wind power in any form, as well natural airflow and ventilation, are considered green. Reducing harmful impacts to the health and welfare of people, flora, and fauna is seen as *green*. Green infers eco-friendly, whereby disposal is not otherwise problematic. Nonetheless, green alone does not carry the far-reaching impact of conservation and sustainability. The need to address the conservation of our resources became apparent in the 1970s when yet another problem surfaced—the uncertain availability of oil.

Instabilities in the Middle East precipitated oil shortages in both 1973 and 1979,[6] limiting availability and spiraling prices that created economic stress and daily inconveniences across the nation. The environmental movement and progressive architects ushered forth environmental and bioclimatic architecture, but it was the sudden instability in our oil supply that added "energy conservation" to the mix. It sparked an interest in alternative energy sources and the need for conservation. Conservation motivated not to reduce emissions, that would take another decade to realize, but *much the opposite*— to keep the turbines running. Nonetheless, the three themes converged: environmentally conscious design; containing pollution; and energy conservation. The concept of sustainable design would soon emerge from this triad, with its special focus on *our common future*.[7]

From Philosophy to Practice

The assertions of humanity's right to a healthy and productive environment, and the threat of human-caused harm, were first addressed on a global scale at the 1972 United Nations Conference on the Human Environment" in Stockholm. Though the word "sustainability" in such context had not yet

---

6.   Arab Oil Embargo of 1973 and oil crises of 1979 due to the Iranian Revolution.
7.   The phrase "our common future" is inspired by WORLD COMMISSION ON ENVIRONMENT AND DEVELOPMENT, OUR COMMON FUTURE (Oxford Univ. Press 1987) [hereinafter OUR COMMON FUTURE].

entered the jargon, humanity's power to alter the human environment took center stage, and so did its man-made component—the built environment.

> Man is both creature and molder of his environment, which gives him physical sustenance and affords him the opportunity for intellectual, moral, social and spiritual growth. In the long and tortuous evolution of the human race on this planet a stage has been reached when, through the rapid acceleration of science and technology, man has acquired the power to transform his environment in countless ways and on an unprecedented scale. Both aspects of man's environment, the natural and the man-made, are essential to his well-being and to the enjoyment of basic human rights—even the right to life itself.

> . . . We see around us growing evidence of man-made harm in many regions of the earth: dangerous levels of pollution in water, air, earth and living beings; major and undesirable disturbances to the ecological balance of the biosphere; destruction and depletion of irreplaceable resources; and gross deficiencies, harmful to the physical, mental and social health of man, in the man-made environment, particularly in the living and working environment.

> . . . Planning must be applied to human settlements and urbanization with a view to avoiding adverse effects on the environment and obtaining maximum social, economic and environmental benefits for all.[8]

Although the corpus of *sustainability* and *sustainable design* had not surfaced by 1972, the *Report of the U.N. Conference on the Human Environment* a half-century ago left no doubt that the human "environment" it addressed incorporated our *human-made contribution*.

The earth's resources and natural environment sustain life and nourish our ability to prosper; they are essential to sustainable development. Sustainability concerns us, our children, and future generations. It reflects a philosophical ideal, that we who occupy this planet have the right to its natural bounty. "We" includes our descendants. We have the rights to fresh air, clean water, and sunlight; the rights to nourishment from the planet's biota and use of its energy—however—we must maintain their viability for *our common future*. Quite simply, we must serve as the stewards. We may use these resources, but not deplete, pollute, or irrevocably poison the earth's biosphere. These elements comprise the essence of sustainable development. They are unlikely to result from happenstance. Consequently, we need a strategy—*sustainable design* is one of the means.

Some of the concepts in sustainable design equate to those employed in environmentally conscious and bioclimatic architecture, but they stem from

---

8.   Report of the United Nations Conference on Human Environment, Stockholm, June 5-16, 1972.

a different root. In sustainable design, the motivation for considering the winter sun, summer shade, and thwarting cold winds is a far cry from constructing a home to be "most pleasant to live in and most useful" as illuminated by Socrates.[9] Nor does sustainable design derive from aligning streets and alleys to exclude the wind as recommended by Vitruvius.[10] It also differs substantially from employing energy-efficient designs solely to lower a building's operating cost. Although these methods will yield beneficial results regardless of the motive, their sole purpose in a sustainable methodology is to provide and perpetuate a healthful environment for the people and organisms that occupy our biosphere—now and in the future. Comfort, efficiency, and cost savings are byproducts, not primary motivations.

Although environmental pollution had captured widespread attention during the 1970s, global warming was just a tiny dot on the radar screen. Public attention focused on industrial pollution, acid rain, and ozone layer depletion. The relationship between carbon emissions and climate change, while undergoing scientific and political debate, did not become an issue in the public arena until it previewed as front-page news in the *New York Times* in 1981.

Study Finds Warming Trend That Could Raise Sea Levels
Walter Sullivan, *New York Times*, Aug. 22, 1981

A team of Federal scientists says it has detected an overall warming trend in the earth's atmosphere extending back to the year 1880. They regard this as evidence of the validity of the "greenhouse" effect, in which increasing amounts of carbon dioxide cause steady temperature increases. . . .

Carbon dioxide in the atmosphere, which is primarily a result of mankind's burning of fuels, is thought to act like the glass of a greenhouse. It absorbs heat radiation from the earth and its atmosphere, heat that otherwise would dissipate into space.

Burning fuels, the connections to transportation exhaust, and the industrial and power sectors' emissions were obvious, but to building design less so. That took another decade to sink in.

The first definition of sustainable development to appear in the context of buildings is credited to the 1987 report issued by the World Commission on Environment and Development, *Our Common Future*. The comprehensive scope of the commission's coverage included long-term strategies for sustain-

---

9.   *See* Xenophon's Memorabilia, *supra* note 1.
10. Vitruvius, The Ten Books of Architecture, *supra* note 2, ch. 2, at 13.

able development and noted that buildings offer significant opportunity for energy savings.[11]

> Sustainable development is development that meets the needs of the present without compromising the ability of future generations to meet their own needs.[12]

Public pressure to reduce environmental pollution and conserve energy brought pivotal changes in the 1990s, and thus, sustainability became a movement.

In 1990, the Building Research Establishment in the United Kingdom published the first environmental assessment tool for new office buildings known as the Building Research Establishment Environmental Assessment Method (BREEAM®). It later expanded to cover a broad variety of buildings and become the first such tool to be used internationally. That year, the American Institute of Architects (AIA) established a Committee on the Environment, funded partly with research grants from the U.S. Environmental Protection Agency (EPA). The extent to which the built environment relates to sustainability through greenhouse gas (GHG) emissions emerged over the next five years under the gazes of the newly founded AIA Committee on the Environment, EPA and the U.S. Green Building Council. Additional credence came from the World Congress of Architects' *Declaration of Interdependence for a Sustainable Future.* In 1990, EPA partially funded the AIA's *Environmental Resource Guide* project with an $800,000 grant.[13]

Published in 1992, the guide contained information on the life-cycle environmental impact of building materials and included case studies. In a memorandum years later, EPA and the AIA proclaimed the guide a "cornerstone in the green building movement."[14] The 1992 U.N. Conference in Rio de Janeiro promoted the free exchange of information regarding the *adverse* environmental and health effects of building materials. Participants proposed the creation of legislation and financial incentives to promote recycling of energy-intensive materials and the conservation of waste energy in building material production. They discouraged the use of construction materials that

---

11. OUR COMMON FUTURE, *supra* note 7, at Chairman's Foreword & 199.
12. *Id.*
13. BuildingGreen News Brief, *Environmental Resource Guide* (Nov. 1, 1992).
14. Memorandum of Understanding between the United States Environmental Protection Agency and the American Institute of Architects (Feb. 10, 2005), https://archive.epa.gov/greenbuilding/web/html/aia-mou.html.

create pollution during their life cycles and recognized the need to increase builder awareness of sustainable technologies.[15]

1993 was a big year for sustainable design. The U.S. Green Building Council was founded, and on Earth Day, President William J. Clinton announced the *Greening of the White House Initiative* to improve the energy and environmental performance of the entire White House complex. The goal included lowering energy use, increasing thermal integrity, employing renewable resources, and improving air quality. Standards for the White House work were based on an AIA-sponsored feasibility study that included design input from more than 90 experts focused on commercially available solutions.[16] At the 1993 World Congress of Architects, the presidents of the International Union of Architects and the AIA, along with 3,000 of their members, signed the *Declaration of Interdependence for a Sustainable Future.* They pledged to "place environmental and social sustainability at the core of our practices and professional responsibilities" and to develop "procedures, products, curricula, services, and standards that will enable the implementation of sustainable design." They also pledged to "educate our fellow professionals, the building industry, clients, students, and the general public about the critical importance and substantial opportunities of sustainable design" and "establish policies, regulations, and practices in government and business that ensure sustainable design becomes normal practice." They vowed to "bring all existing and future elements of the built environment—in their design, production, use, and eventual reuse—up to sustainable design standards." The imperative to implement sustainable design could not have been clearer, an item to be reckoned with on the agenda. By the end of the decade, the U.S. Green Building Council (USGBC) released the pilot program for its building certification system named Leadership in Energy and Environmental Design, referred to as LEED®, and the first 12 candidate buildings were certified in 2000.[17]

Rachel Carson campaigned against pollution in the 1960s. Calls for energy conservation arose from oil shortages in the 1970s. The possibility of climate change emerged in the 1980s, and the building industry awakened to the call in the 1990s. All summoned society to a common purpose—the viability of our future. By the turn of the millennium, a methodology for

15.   United Nations Conference on Environment & Development, AGENDA 21, Rio de Janeiro, Brazil, June 3-14, 1992.
16.   U.S. DOE, Office of Energy Efficiency & Renewable Energy, Greening of the White House, Six Year Report (DOE/EE-000) (Nov. 1999).
17.   LEED, *More Than a Decade of High-Performing Buildings,* Feb. 21, 2013, https://www.usgbc.org/articles/more-decade-high-performing-buildings-out-now-edcs-february-issue.

sustainable design was fully developed, percolating throughout the design-build industry. The apparatus to make significant strides toward a sustainable future was in place. Environmental and sustainable design progressed from the environmental challenges of early human survival to the existential threat of global warming, from human instinct to modern science, from philosophy to practice.

The 21st Century

The forces aligned in the 21st century's first decade to build momentum in the United States. Architects, planners, educators, federal agencies, and industry united in support of green building, sustainable design, and sustainability. Solar electric, solar heating, geothermal heating, hydrothermal cooling, and wind power were maturing to a commercial reality. By 2005, the U.S. government had taken a more active role. EPA and the AIA signed a Memorandum of Understanding[18] to advance their informal working relationship, another step forward signifying the importance of sustainable design.

> EPA's mission, to protect human health and the environment, and the AIA's mission, to improve the quality of the built environment, come together in the emerging fields of sustainable development and green building. These two organizations are therefore working together to ensure that we build in ways that enhance and regenerate Nature, while providing vibrant, healthy places for people to live, work, and play. In essence, EPA and the AIA are uniting in the goal of promoting development that sustains the environment.

> Design, construction, and development . . . have a significant impact on the environment at every level—locally, regionally, nationally, and globally. These impacts occur during all building stages—from the extraction and manufacturing of building products through siting, design, construction, operation, maintenance, renovation, and ultimate removal of the building and its components.

No waffling here, the direct tie between the built environment and its potential for environmental damage was fully acknowledged:

> The built environment also can adversely affect the natural environment through air and water pollution, solid and hazardous waste generation and disruption of wildlife habitat, the hydrologic cycle, and the climate.

---

18. Memorandum of Understanding between the United States Environmental Protection Agency and the American Institute of Architects (Feb. 10, 2005), https://archive.epa.gov/greenbuilding/web/html/aia-mou.html.

All the pieces lay in place for the widespread implementation of sustainable practices in building design and construction except one: fast, accessible, and affordable data analysis. In other words, affordable people-friendly computers with design and analysis application programs for architects, planners, and designers. That final piece, the key to its general application by architecture firms of all sizes, became available commercially mid-decade. Building science and design techniques, now aided by personal computers and dedicated computer programs, spawned the proliferation of sustainable design for everything from houses to tall buildings to infrastructure. Many of these programs were adapted from their aerospace and high technology counterparts.

While the architecture and planning professions judiciously assimilated these new tools and methodologies, they flourished quickly in the academic sector too, where experimentation became a driving force for undergraduate and graduate college programs. Aided by the AIA, the U.S. Green Building Council, and many other organizations and industries, the effort to disseminate this know-how filtered throughout the industry. The design potential was and remains astounding. The dream of designing a green and sustainable built environment had the potential to become a reality. The AIA launched a movement to permanently embed sustainable design within the practice of architecture in 2009. As the "issue of climate change and the impact of buildings on carbon emissions" required architects to find solutions that leave "greener footprints," and recognizing that "[s]ustainable design has evolved from a niche service offering to a profession-wide imperative," the AIA's Board of Directors made *sustainable design* education courses a mandatory annual requirement for architects to maintain their membership.[19] This immediately impacted its members. Even the *New York Times* noted this phenomenon:

Architects Return to Class as Green Design Advances
Robin Pogrebin, *New York Times*, Aug. 19, 2009

It seems like only yesterday that environmentally conscious building practices began making their way into the architecture profession. How times have changed. This year, the American Institute of Architects implemented a policy requiring all members to take four hours of continuing education courses in sustainable design every year. The requirement, which extends through 2012, represents a response to a rapidly changing field and a recognition that architects must continue to refresh their knowledge of sustainable construction methods and building materials.

19. AIA, Continuing Education System Survival Guide v7 (2009).

. . . Licensed architects learn about subjects like building form, or how the shape of a building responds to the environment; energy modeling, including how much energy it takes to operate a building and ways to reduce the carbon footprint; how to reduce heat gain from sunlight; the most energy-efficient ways to position buildings relative to the sun, wind and other elements; ways to bring in natural light and reducing electricity consumption; and the preservation and reuse of existing buildings.

. . . This kind of expertise is now being applied to every aspect of design and construction, experts say, from how materials are transported to and disposed of at a work site, to the tools and machines used, to consideration of how a building will perform over the next half-century.

AIA's membership exceeded 94,000 by 2018.[20]

Looking back over the last 50 years, anyone born after 1970 in the United States, Canada, western Europe, Australia, Japan, or in many other locations worldwide, came of age in a country working to achieve a green and sustainable future. If schooled in architecture or planning, husbandry or ecology, design or engineering, one's education included "sustainability" and "green." If born after 1985, they were embedded in your curriculum. "Sustainability," "sustainable design," and "green" became common jargon. From 1988 to 2008, use of the phrase "sustainable design" increased 125 fold in Google's database of over five million books.[21] This is not to imply unanimity over the cause of changing climates or that every polity is willing to minimize the use of fossil fuels regardless of its constituents' needs. Nonetheless, more than 200,000 architects, engineers, and designers had attained proficiency in sustainable design, construction, and operating standards, qualifying as a U.S. Green Building Council LEED® Professional. It took 50 years of caring and environmental activism to get to this point, the last 20 making tremendous strides. By 2019 the U.S. Green Building Council was LEED® certifying more than 2.2 million built square feet (200,000 m²) each day,[22] having registered or certified more than 1.5 million residential units and 100,000 commercial and government projects.[23] Commercial and institutional property

---

20. AIA, AIA Year in Review 2018 (2019).
21. Google NGram data 1988 to 2008.
22. Press Release, U.S. Green Building Council, Green Building Accelerates Around the World, Poised for Strong Growth by 2021 (Nov. 13, 2018).
23. Press Release, U.S. Green Building Council, U.S. Green Building Council Announces Annual LEED Homes Awards, Recognizing Residential LEED Projects Elevating the Living Standard Through Sustainable Design (June 20, 2019); Selina Holmes, U.S. Green Building Council, *Infographic: LEED Reaches Over 100,000 Commercial Projects* (Nov. 7, 2019).

certifications alone exceed 6.5 billion square feet (600 million m²).[24] The U.K. BREEAM® program has registered or certified more than 2.5 million buildings in 70 countries, of which 118 million square feet (11 million m²) were certified as of 2016.[25] These are impressive numbers. The AIA and the American Planning Association (APA) have made sustainable design of the built environment a core principle for more than 135,000 professionals.[26] Public education through activism has also made significant strides. As of January 2021, The Climate Reality Project has trained over 31,000 climate change activists to mobilize communities in 170 countries. More than 190 countries celebrate Earth Day with annual activities. Architecture, planning and engineering schools have integrated sustainability and green programs in their curriculums. With sustainable principles embedded in the educational system, trained professionals, public suasion, and new tools for environmental design, the United States saw nearly 30 billion square feet of added construction from 2005 to 2017 with building sector carbon emissions decreasing by 20%. What an auspicious inauguration for this new millennium. We are well on the way to make "green" and "sustainable design" the norm. Yet, there is something wrong with this picture, something seems wrong with the outcome.

I would happily end this book here if this 20% reduction in carbon emissions were due primarily to building design and construction. But it was not. Some of those decreases in carbon emissions have been attributed to the impact of the economic downturn; some from abandoning coal as a fuel source in favor of natural gas; some from switching to Compact Fluorescent and light-emitting diode (LED) lighting; some from installing more efficient appliances; and some due to the weather. Reductions in carbon emissions derived primarily from the composition of the energy purchased for building "operations," lower heating and cooling requirements due to milder weather, and the installation of compact fluorescent and LED lamps add to the list. They related mostly to the purchase of electricity, with reductions credited to the declining use of coal for electricity generation, "decarbonization," not to architects' building designs nor planning or policy. Total electricity use fell slightly over those 12 years even though the U.S. building stock increased significantly, but not due to building design. The weather strongly influ-

24. GRESB B.V., subsidiary of Green Business Certification Inc., *U.S. Green Building Council Releases Annual Top 10 States for LEED Green Building Per Capita* (Feb. 2, 2018).
25. Eleni Soulti & David Leonard, BRE Global Ltd., U.K., The Value of BREEAM: A Review of Latest Thinking in the Commercial Building Sector (2016) and BRE Global Ltd., U.K., BREEAM In-Use Fact File 2017 (2016).
26. AIA 2018: 94,000 members; APA 2018: 43,000 members.

enced electricity use from the sector's residential component. In 2017 and the preceding six years, heating and cooling degree days held below the 10-year average by as much as 15% during the heating season and 2% during the cooling season.[27] Energy use reductions due to lighting efficiencies improved dramatically with the introduction of compact fluorescent and LED lamps, with the residential LED market share exceeding 33% in 2017, their luminous efficacy being seven to eight times more efficient than incandescent lamps, which decreased to less than 5% of the market.[28] Coal-fired electricity in 2017 decreased to 60% of its 2005 level.[29] Carbon emissions dropped 28% primarily due to the declining use of coal.[30] Approximately three-fifths of that reduction was attributed to the switch to natural gas and two-fifths to the use of zero-carbon electricity generation such as wind and solar.[31]

Neither more efficient building designs nor reducing "embodied" emissions triggered these improvements. Absent the introduction of compact fluorescents and LEDs, the switch to natural gas and the mild weather, the picture would have been grim. The sustainability agenda had been set: the built environment's role, the imperative for sustainable design, and the need to address cause and effect. Yet, when it came to the buildings themselves—building design and construction—sustainability was not attained, it was little more than a label, a palliative we will grow to regret. Though the industry possesses the savvy to tackle the problem, it had ignored the fundamentals even though they set the stage for the century to come. We focused on future gains while ignoring the present. After 50 years of science, engineering, education, and policy we learned how to "talk the talk"—but did we "walk the walk"? While material processing and fabrication generate high levels of emissions that are built into a building's design and construction methodology, we addressed construction expediency at the expense of sustainable design. We continued to rely on the projection of future savings rather than the current conservation of carbon emissions. We hang our hope on a carbon-free energy future that will take 30 to 60 years to achieve on a global scale. Unfortunately, the impact of carbon-induced global warming will not wait.

The construction of every building and element of infrastructure affects the long-term viability of the environment from the moment raw materials are mined or harvested. That revelation, relatively new in the history of archi-

---

27. U.S. Energy Info. Admin., Energy-Related Carbon Dioxide Emissions (2017) [hereinafter Energy-Related Carbon Dioxide Emissions].
28. United Nations Environment Programme, 2018 U.N. Global Status Report (2019).
29. U.S. Energy Info. Admin., Annual Energy Outlook (2019).
30. Energy-Related Carbon Dioxide Emissions, *supra* note 27.
31. *Id.*

tecture and planning, surfaced for the first time within just the past several decades. The reality of cause and effect and the existential role of design have developed slowly, despite their being crucial to the sustainability of our habitat. We have yet to absorb the extent of this relationship: tying design to the welfare of future generations; relaying meaningful information to those who design, implement, and regulate the built environment. More than 30 years into the environmental design learning curve we still struggle with its effective application.

Chapter 3

Parsing Carbon, Sustainability, and Green

Oddly enough, though we strive to make carbon-based energy obsolete, our "green" environment relies upon the element carbon, relying on a delicate balance in its ebb and flow throughout the earth's biosphere. Carbon is essential to life, environment altering in its excess, and deadly in some compositions. A key component of all life on earth, carbon constitutes 18.5% of human body mass, and almost half of the earth's dry biomass.[1] By mass, carbon is the fourth most abundant element in the universe. We cannot exist or survive without it, yet it can undo the chemistry of our habitat and our very existence. The earth's carbon atoms are as old as the planet, part of its structure, and integral to the code of life's DNA. Combined in counterintuitive extremes with a gamut of attributes, they form both substances lacking shape or volume and solids of extraordinary strength: gases, liquids, soft solids, and rock-hard crystals such as diamonds. These can be clear or invisible, reflect or pass light, or absorb all wavelengths to appear black. Alone or in combination with other elements, carbon's bonds can provide physical structure or form catalyzing compounds that liberate molecules and discharge energy. Carbon can free up oxygen or render it poisonous, or be activated to filter and cleanse. It can nourish air, purify water, fuel civilizations, or pollute the environment and alter the climate. Metaphorically, carbon can be *black* or *green*.

Carbon's natural cycle through the earth's geosphere—the lithosphere, hydrosphere, biosphere, and atmosphere—changed very slowly over millions of years to become favorable to human habitation. Nevertheless, over a relatively short time frame, industrialization and carbon resource reliance have put that cycle on a fast track to destabilization catalyzed by their byproduct

---

1. Food & Agriculture Organization of the United Nations, *Knowledge Reference for National Forest Assessments—Modeling for Estimation and Monitoring*, http://www.fao.org/forestry/17111/en/ (last visited May 6, 2021).

carbon dioxide (CO_2). Although it is the fuel that drives photosynthesis liberating oxygen to breathe, carbon green—in overabundance, CO_2's influence is *black*.

CO_2 is a natural component in the life cycle of terrestrial and aquatic animals and plants. That includes humans. Simply put, animal life ingests oxygen and expels CO_2 while plants do the reverse. Animal life uses the oxygen to process sugars, releasing energy and CO_2. Plants thrive on the opposite. Absorbing carbon dioxide and the sun's energy for photosynthesis, they produce sugars for growth and release oxygen to the atmosphere. This synergy between animal and plant life utilizes the atmosphere and water-bodies as its medium. When other CO_2 emissions overwhelm the balance, however, things go amiss. Over time, industrialization, increased population, and the collective impact of human activity have upset the natural balance. It remains to us to restore and maintain that balance, to keep carbon *green*.

Of all the greenhouse gas (GHG) emissions capable of altering the earth's temperature and thereby its climate, CO_2 has been the most significant destabilizer in modern times. Methane has a more potent impact, but fortunately a lower volume of emissions. GHGs regulate the earth's temperature by absorbing some of its infrared heat energy and re-emitting part, warming the atmosphere and the surface. Constancy in this activity is crucial to maintain a temperature range fit for human survival. Water vapor is also a GHG. Combined with stratiform clouds in the atmosphere, the two contribute nearly three-quarters of the greenhouse *heat re-radiation*. CO_2 represents just 20% of the GHG mix, but its enduring presence influences air temperature, thereby its water vapor moisture content. Changes in atmospheric temperature and pressure cause water vapor and clouds to precipitate rain, sleet, and snow—but unlike water vapor, CO_2 neither condenses nor precipitates from the atmosphere. Though only 20% of the mix, its overall greenhouse effect amounts to 80% by regulating the atmosphere's water vapor content.[2] The atmosphere cools when the concentration of CO_2 falls and warms when it rises. When it cools, water vapor condenses out of the atmosphere, but rising temperatures cause water to evaporate off the earth's surfaces; those vapors amplify the greenhouse effect thus warming the planet.

In 2019, approximately 70% of *all global emissions* resulted from fossil fuel consumption, 38 billion metric tons of fossil CO_2 were pumped into the atmosphere worldwide, more than 6 billion metric tons[3] generated by the

---

2. Andrew A. Lacis et al., *Atmospheric CO_2: Principal Control Knob Governing Earth's Temperature*, SCIENCE, Oct. 15, 2010.
3. U.N. ENVIRONMENT PROGRAMME, U.N. EMISSIONS GAP REPORT 2020 (Nov.. 2020); U.N. ENVIRONMENT PROGRAMME, U.N. EMISSIONS GAP REPORT 2019 (Nov. 2019). Reference for U.S. emissions

United States. This equates to burning 600 billion gallons of gasoline or 6 trillion pounds of coal in the United States; six times that worldwide.[4] Nearly 40% of those emissions occurred from the construction and operation of buildings alone. *The built environment is our largest source of energy-based climate changing emissions, traceable to the design and operation of our buildings and infrastructure. Consequences from what we build today will persist for the next 50 to 100 years.*

Fortunately, the oceans, lakes, rivers, forests, urban trees, and vegetation absorb about half of the CO_2 emitted each year. Some of the balance will be resolved through the natural carbon cycle over the long term, though at a level and rate insufficient to keep our biosphere in balance. Twenty percent of the emissions could linger in the atmosphere for a millennium.[5] The earth, oceans, forests, and other biomasses are banks that sequester carbon, much of it for the long duration. But as oceans approach their absorption capacity they suffer acidification; with an overabundance of CO_2, vegetation overgrows due to the warming climate. Such changes unleash other undesirable stimuli. Eventually these excesses will recycle back—but in millenniums not lifetimes. Given the magnitude of emissions attributable to the design and construction of our buildings and infrastructure, architects, engineers, planners, and policymakers can substantially reverse this trend through materials selection and use. Thankfully, we understand the cause and effect. Thankfully, we can effect a change through "green" design, or is it by designing sustainably? Although we often use the terms interchangeably, their scope is not the same.

Sustainable or Green?

While closely related, "green" and "sustainable" are not equivalent. Sustainability is "green," but green practices are not necessarily "sustainable." Green implies that a practice or a product's use or manufacture is benign to people and the environment. Green implies that something is healthy and eco-friendly for its particular function, as does sustainable. Nevertheless, a sustainable design embraces its overall ramifications—*their enduring reach.* Sustainable, more heroic, addresses the earth's future habitability and the course of human life. Sustainable addresses the mechanisms and immediacy

statistics: U.S. ENVIRONMENTAL PROTECTION AGENCY (EPA), INVENTORY OF U.S. GREENHOUSE GAS EMISSIONS AND SINKS 1990-2016 (2018) (EPA 430-R-18-003).

4.   U.S. Environmental Protection Agency, *Greenhouse Gas Equivalencies Calculator*, https://www.epa.gov/energy/greenhouse-gas-equivalencies-calculator (last visited May 6, 2021).

5.   NASA Earth Observatory, *The Carbon Cycle*, https://earthobservatory.nasa.gov/features/Carbon Cycle/page1.php (last visited May 6, 2021).

of global warming. Green and sustainable have become common words in the lexicons of environmentalism, architecture, and planning. Highly overused, they are more often bandied about for commercial sales pitches than for meaningful design. As feel-good terms, marketers appropriated both for their appeal and selling power, helping to promote everything from cosmetics to office supplies and places to live, work, or vacation. It is no wonder that in a 2017 survey designed to mirror the composition of the U.S. population, 70% of the respondents rated the word "sustainable" as a positive.[6] Equated to social responsibility, buying "green" or "sustainable" extrapolates to doing one's part. Unfortunately, when an action or product, or even an entire building is promoted as "green" or "sustainable," the public *presumes* it to be so—without access to the facts, without scrutiny. Such labels satisfy our call for environmental action, to do one's part fighting poisons, pollution, and global warming. This human folly is more insidious than the Emperor's New Clothes exposed by a child's uninhibited observation; here the naked facts are not apparent—our praise is ignorantly sincere.

Sustainable and green are frequently used interchangeably, sometimes in the same sentence or paragraph to cover all bases. This real and widespread confusion infects not only the public discourse but also those responsible for its environmental education and protection. Even the U.S. Environmental Protection Agency's (EPA's) website seems confused, where green and sustainable are defined with elements of each other, as one in the same. No clarity there either.

> Green, or sustainable, building is the practice of creating and using healthier and more resource-efficient models of construction, renovation, operation, maintenance and demolition.[7]

> Green building is the practice of creating structures and using processes that are environmentally responsible and resource-efficient throughout a building's life-cycle from siting to design, construction, operation, maintenance, renovation and deconstruction. . . . Green building is also known as a sustainable or high performance building.[8]

On EPA's *Home* page, "Green building" is equated to "sustainable building," both in terms of health and resource efficiency thereby covering all the bases. The *About* page more aptly defines "sustainable building," but it notes that "[g]reen building is also known as a sustainable or high performance building," high performance referring to the attributes of a sustainable build-

---

6.   *Eco Pulse™ 2017 Special Report, United We Stand* (Shelton Group Inc.).
7.   U.S. EPA, *Green Building* (updated Feb. 20, 2016), https://archive.epa.gov/greenbuilding/web/html.
8.   *Id.*

ing—confusing, take your pick. If it is labeled green it must be sustainable, which means it is high performance.

Why the lack of clarity? *Green* is a simple concept, eco-friendly, benign to people and the environment. *Sustainability* requires a strong commitment to future generations, to resource conservation. *Sustainable design should be a science*—not a marketing tool, feel-good label, or science fiction. A built environment designed to be "environmentally responsible" and "resource efficient" through its entire *life cycle* constitutes sustainability. This means responsible and resource efficient from conception through construction, operation, maintenance, renovation, and eventual deconstruction or demolition. It includes the life cycle of its primary materials and the overall efficacy of the design—the entire system assembly not just a single component or product. It must meet present needs without compromising the ability of future generations to meet their own. Sustainable design is an operative process with infinite reach—*future needs* are the qualifier. Sustainability is not simply green. Does that matter? Yes. Sustainability requires substantial forethought and planning. It is neither simply a label nor a collection of points in a rating system. If it is a choice between sustainable or green, choose sustainable!

Energy and Carbon

A building's consumption of *operating energy* is easily understood and convenient to monitor by tracking the purchase of electricity and primary fuels. Because operating energy's carbon emissions recur throughout a building's lifetime, operating energy is usually highlighted as the earmark of sustainability—a focal point for reducing or *offsetting carbon emissions*. Reducing operating energy is a policy objective in many countries and municipalities around the world. Regardless of a building's inherent design efficacy, taking advantage of equipment efficiencies, user efficiencies, and onsite energy capture will significantly reduce net operating needs and thereby operating emissions. Installing energy-efficient appliances, which is typically a matter of budget or availability, is not a challenge of building design. Neither is instilling a culture of energy-efficient occupant behavior, transferring energy conserving responsibilities to the occupants. Switching to low-carbon or no-carbon fuels where available directly addresses the root of the problem, but rarely taxes a design nor the need for operating efficiency. Nonetheless, over time, all of these measures will retard future increases in global warming to some degree. It is no surprise that installing energy-efficient appliances, instilling energy-efficient user habits, or switching to low- or no-carbon fuels are the most prominent "go-to" means in the methodologies of sustainable design, but these are the low-hanging fruits. The use of low-carbon or no-carbon fuels by its very nature reduces operating emissions, though the large-scale elimination of fossil and biofuels requires participation on a political, industrial, and global scale. Ultimately, carbon-free energy will be the solve-all, but waiting for that to happen soon enough is too risky a bet.

On the other hand, onsite energy capture or generation and "net-zero" energy design fall in a different category; all are tied to a facility's design. To be designated a zero energy building, onsite energy capture and "renewable" energy generation must at least balance the energy consumption. Zero energy buildings are hard to achieve, requiring significant engineering precision and architectural finesse. In addition to a low energy loss building design, install-ing energy-efficient appliances and generating energy onsite, success depends upon energy-efficient occupant behavior as well. Though sustainably ben-eficial in that they generate *renewable* energy onsite, *zero energy* does not mean "zero" nor even "net-zero" operating carbon emissions. The common definition of a zero energy building just requires the onsite generation of *renewable* energy, such as from solar radiation, wind, water flow, the earth's temperature differentials, or biological processes. *Renewable* energy means sourced from natural energy banks that are constantly replenished on a human timescale, but it need not be carbon free—biofuels qualify. Nor must it be "used" onsite to qualify as a zero energy building. The building may buy or use energy from any external source—*including fossil fuels*—as long as it exports an equal or greater amount of *renewable* energy that has been gener-ated onsite.[9] If all of the *renewable* energy generated onsite were to be uti-lized, perhaps from storage, it would improve the carbon mix. Unfortunately, that is unlikely to happen. Current storage capacities are insufficient to hold excess generation, and many states in the United States cap the total renew-able energy allowed into the grid from private and government installations. With the uncertainty of balancing fossil energy with renewable generation that is useful to the grid, the sanction of biofuel emissions, and the stringent maintenance required over the long term, zero energy buildings do not guar-antee significant lifetime reductions in carbon emissions.

The efficiency of appliances, lighting, controls, and accessories clearly play a very important role in reducing operating energy. But if a building were to require less heating or cooling by design, less energy would be consumed to heat or cool it in the first place, and less heating or cooling capacity would

---

9.   National Inst. of Bldg. Sciences, U.S. Dep't of Energy, *A Common Definition for Zero Energy Buildings* (Sept. 2015).

need to be installed. The same applies to providing sufficient insulation, appropriate facade and room orientations for natural light, and screening from direct solar radiation. Such design elements would lower the need for heating, cooling, and lighting—thereby lowering operating energy and carbon emissions—reductions resulting solely from thoughtful building design, regardless of the efficiency of appliances, equipment, lighting, or a low-carbon energy source. Although these design-based efficiencies will minimally impact plug loads, cooking, refrigeration, and other appliances, the design and orientations of the building envelope can significantly reduce the energy requirements for lighting, heating, cooling, humidifying, and ventilation. This too must be handled with architectural finesse and engineering precision, however, as rarely do material and physical solutions arrive carbon free. A building's materials and construction come with their own carbon profile—their own footprint of *embodied energy* and its companion embodied carbon.

"Embodied" Is Nothing to Brag About—Not for Energy, Not for Carbon

Although one's personal quest to embody energy might seem auspicious, embodied energy and carbon are quite the opposite in the physical world: it indicates energy already *exhausted*, carbon already *released*. *Embodied* does not imply a *usable store of energy*, nor carbon encapsulated safe from future emission. Embodied energy is a measure of all the energy spent to extract, process, manufacture, transport, install, and dispose the materials and assemblies that constitute a building, which vary with the type of construction. Embodied carbon represents a measure of all the carbon emissions released during that entire life cycle—from cradle to grave. Or more optimistically, from cradle to re-use. For example, embodied carbon includes the emissions released from the energy spent to mine or harvest raw materials, to process and manufacture the product, transportation from site to site, and assembly. It also includes the emissions associated with the energy consumed during the building's assembly and construction process. But this is merely the initial embodied carbon. Once a building is occupied, its repair, maintenance, and renovation continue to add to the embodied carbon tally, to be followed by energy-related emissions during its eventual demolition and disposal. A structure's embodied energy indicates intensity—how much energy was required to produce each component and ultimately the life cycle of the entire edifice. Embodied carbon indicates the carbon intensity of its footprint, how much CO_2 was and will be released in the expenditure of the embodied energy. Such knowledge allows for energy conservation by intelligent choice, by function of design. Once occupied, a building's *operating energy* comes into play. Operating energy and carbon, however, are the second part of the total equation, an expression of the energy that *will be* consumed and the carbon that *will be emitted* over a building's useful life. Their magnitude relates not only to building design, but also relies heavily on appliance efficiencies, the occupants' functional choices, and good practices as mentioned above. Moreover, circumstances such as the weather and the economy play significant roles, as do technical advances such as the gradual installation of energy-efficient appliances, light-emitting diodes (LED) lighting, and the like. It is important to distinguish sound design and construction practices from the apparent gains derived from climate and economic anomalies and trends. Statistics without sufficient scrutiny often tout misleading praise or sounds of alarm. Operating carbon contains unpredictable variables and can increase or decrease over time. On the other hand, embodied carbon is *already emitted*, already *baked-in*, with some yet to be released during future maintenance, replacement, and eventual deconstruction.

A Building's Embodied Carbon

For more than 30 years, the architecture, engineering, and construction industry has studied the relationship between fossil fuel energy and its resulting GHG emissions. Those studies led to a better understanding of the underlying problems in constructing an environmentally sound built environment. By differentiating between *operating energy* and *embodied energy*, architects, engineers, and manufacturers have responded with an ever increasing variety of materials, technologies, and design techniques to reduce our energy needs and lower the energy embodied within a building's components. Carbon emissions from building operations play a significant role as they occur continually over a building's lifetime, but *embodied carbon reflects the more immediate danger*—the damage already incurred due to the architectural design—the material choices and construction methodology. Most of the damage arises from emissions released to the atmosphere before occupancy. They are impacting the carbon cycle's warming mechanism now, exacerbating conditions that may cause warming to irreversibly tip, and will continue to do so for a century or more. I mention this not to cry over spilt milk, but to highlight a recurring problem triggered with each order for building materials, for every building now in construction, or on the drawing board for future construction. Every delivery of building materials from stock or manufactured for stock replenishment has already added its emissions to the atmosphere—before building completion, commencing with raw materials processing.

We cannot ignore a building's history, its DNA. Every material, every component, and every construction operation has its own history of energy expended and GHGs emitted. The mere existence of a physical component translates to carbon already released and a portion remaining for release in the future. None of those emissions get a free pass. They have already contributed to the global warming process and will do so again at the end of each component's useful life. Choices made by parties up and down the line matter: from mining to manufacture, from packaging to transportation. Whether a material or component is selected to conform to the building code, to satisfy an architect's design, an engineer's specification, or a client's preference, it is chosen from multiple options—there is a choice. That choice matters. Its carbon footprint and future emissions are already built in. Order it and more of the same will be fabricated whether a low-carbon product or high. As both operations-driven emissions and embodied emissions are innate to a building's design, materials, and construction, sustainability begins by minimizing embodied carbon through materials selection, and operating carbon through environmental design. This includes equipment, appliances, and lighting which are frequently ignored when tallying embodied carbon. They too embody carbon during manufacture, maintenance, and disposal. Though a complex compilation that can be difficult to conjure, *embodied carbon is the first measure of sustainability* when faced with a near-term time frame. As the bulk of those emissions occur in surges early on, they are a leading indicator for immediate environmental impact; increasing the *current* concentration of atmospheric carbon, they accelerate the rate of emissions-induced warming.

Chapter 4

Reality and the Call for Action

Global warming is not a function of the energy we consume,
but rather the release of its carbon content.

The Reality

More than 25 million metric tons of carbon emissions are released to the atmosphere on a daily basis in conjunction with building operations.[1] This is simply to maintain the comfort and functionality of our living and working environments. Lighting, heating, cooling, humidifying, and ventilating are significant sources of those emissions along with plug loads, cooking, and refrigeration. The complete list is long. Some of these uses emit their greenhouse gases (GHGs) directly to the atmosphere, such as when burning natural gas or other fuels for heating or cooking. Others emit them indirectly via the electricity they consume, which is generated by burning natural gas, oil, coal, or other fuels. Emissions from industrial facilities and transportation round out the list. In 2019, fossil-related carbon emissions from energy and industry reached an historic one-year high of 38 gigatonnes (Gt) worldwide. This helped drive the atmospheric concentration of carbon dioxide (CO_2) in 2020 to its 10th consecutive ½% single-year increase since 2010. Ten Gt were from building operating emissions, which have increased more than 2% for the third consecutive year. A review of past levels provides perspective for the accelerating rate in recent decades. Ice core samples reveal that prior to the 20th century, CO_2 concentrations last reached 300 parts per million (ppm) more than 300,000 years ago and had not exceeded that level for 800,000

1.   Data and computations are from INTERNATIONAL ENERGY AGENCY (IEA) & UNITED NATIONS ENVIRONMENT PROGRAMME (UNEP), GLOBAL STATUS REPORT FOR BUILDINGS AND CONSTRUCTION: TOWARDS A ZERO-EMISSIONS, EFFICIENT, AND RESILIENT BUILDINGS AND CONSTRUCTION SECTOR 2018 (2018); IEA & UNEP, GLOBAL STATUS REPORT FOR BUILDINGS AND CONSTRUCTION: TOWARDS A ZERO-EMISSIONS, EFFICIENT, AND RESILIENT BUILDINGS AND CONSTRUCTION SECTOR 2019 (2019); and IEA & UNEP, GLOBAL STATUS REPORT FOR BUILDINGS AND CONSTRUCTION: TOWARDS A ZERO-EMISSIONS, EFFICIENT, AND RESILIENT BUILDINGS AND CONSTRUCTION SECTOR 2020 (2020) [hereinafter GLOBAL STATUS REPORT FOR 2018; GLOBAL STATUS REPORT FOR 2019; and GLOBAL STATUS REPORT for 2020].

years. After reaching 300 ppm in 1910, the level increased during the next 50 years by merely 17ppm to 317ppm in 1960. Then the climb accelerated: 50 ppm in 39 years to 367 ppm in 1999; then another 50 ppm to a monthly average record of 417 ppm in only 21 years in May 2020.[2] The last time our planet reached these concentrations was more than three million years ago when temperatures were 2 degrees Celsius (°C) to 3°C above pre-industrial levels, with sea levels 50 to 80 feet (15-25 meter (m)) higher than today.[3] At the current rate of increase, our atmospheric CO_2 concentration will climb by more than 5% per decade. "CO_2 emissions increased by nearly 0.5% for every 1% gain in global economic output"[4] and the "average global surface temperature rise is almost linear with cumulative CO_2 emissions."[5] This is not an encouraging picture.

The 2015 Paris Agreement strove to stabilize GHG emissions as soon as *possible* followed by further rapid reductions, to limit the globe's average temperature rise to "well below" 2°C. Limiting the increase to 1.5°C by balancing emissions and carbon gas removal in the second half of this century was the long-term goal. Nevertheless, when the initial *voluntary* commitments from the signatories were submitted and mid-century targets were issued, the International Energy Agency (IEA)[6] adopted an emissions reduction protocol aiming at a less stringent 2°C limit by 2100 known as the *2DS Scenario*. This goal sought to limit GHG emissions from 2015 through 2060 to 1,000 Gt cumulatively, with a mere 170 Gt allowed for the remainder of the century.[7] Yet, even if this less stringent target would be achieved, it had only a 50/50 chance of capping the average global temperature increase to 2°C—not very favorable odds. Just two years later, with global emissions already exceeding 32 Gt per year since 2015 and unlikely to level off until mid-century or later, the IEA revised its position. They concluded that current commitments and

2.   NOAA Climate.gov, based on EPICA Dome C data (D. Lüthi et al., 2008) provided by the National Oceanic and Atmospheric Administration (NOAA) and the National Centers for Environmental Information (NCEI) Paleoclimatology Program. *See* NOAA, *Rise of Carbon Dioxide Unabated*, June 4, 2020, https://research.noaa.gov/article/ArtMID/587/ArticleID/2636/Rise-of-carbon-dioxide-unabated; Ed Dlugokencky & Pieter Tans, NOAA/ESRL, *Earth System Research Laboratories' Global Monitoring Laboratory of the National Oceanic and Atmospheric Administration*, www.esrl.noaa.gov/gmd/ccgg/trends; NASA Goddard Institute for Space Studies, Ice Core Data Adjusted for Global Mean, *Forcings in GISS Climate Mode, Well-Mixed Greenhouse Gases, Historical Data*, https://data.giss.nasa.gov/modelforce/ghgases/Fig1A.ext.txt (data for 1910 and 1960).
3.   Climate.Gov, *Climate Change: Atmospheric Carbon Dioxide*, Aug. 14, 2020, https://www.climate.gov/news-features/understanding-climate/climate-change-atmospheric-carbon-dioxide.
4.   Statistics reported in IEA, Global Energy & CO₂ Status Report: The Latest Trends in Energy and Emissions in 2018, 7 (Mar. 26, 2019).
5.   IEA, Energy Technology Perspectives (2017) [hereinafter Energy Technology Perspectives].
6.   The IEA is an autonomous intergovernmental organization made up of 30 Member and 8 associate countries.
7.   1 Gt = 1 billion metric tons.

practices were not sufficient, and an increase of 2.7°C by 2100 was more likely with continuing increases thereafter.[8] They also concluded that the built environment's emissions were a significant part of the problem: "The vast majority of countries today have buildings-related carbon intensities . . . far from the CO_2 intensities required to meet ambitions for 2°C or below."[9] The *IEA Global Status Report* for 2017 commenced with this warning: "The global buildings sector is growing at unprecedented rates, and it will continue to do so. Over the next 40 years, the world is expected to add 230 billion square metres in new construction—adding the equivalent of Paris to the planet *every single week*." This was further equated to "building the current floor area of Japan every single year from now until 2060." The then current trajectory would add "as much as 415 gigatonnes of CO_2 to the atmosphere—half of the remaining 2°C carbon budget and twice the total building emissions from 1990 through 2016."[10] "While buildings sector energy intensity has improved in recent years, this has not been enough to offset rising energy demand. Buildings-related CO_2 emissions have continued to rise by around 1% per year since 2010." The report concluded: "Building energy certification is inadequate to influence major change in the buildings market, even if it is becoming increasingly common (although it is typically voluntary or covers only a certain number of buildings)."

In October 2018, the Intergovernmental Panel on Climate Change (IPCC), the United Nations' (U.N.'s) body for assessing the science of climate change, issued a more realistic assessment of the measures required. It found that in order to contain warming to the ultimate target of 1.5°C, human-caused CO_2 emissions must be roughly halved by 2030 from the 2010 level, and near net-zero around 2050.[11] The report, prepared by 91 authors and editors from 40 countries, cited more than 6,000 scientific references with contributions from "thousands of expert and government reviewers worldwide." While getting to net-zero is a long way off, our achievements over the next 10 to 15 years are key to stabilizing annual emissions, the critical milestone from which we can start reduction. A good part of that effort will be determined by the carbon intensity of the new buildings' construction and existing building renovations. Approximately half of the new construction projected for 2020 through 2050 will be completed by 2035. Half of the embodied emissions that must be curtailed to meet the 2050 target,

---

8. *See* ENERGY TECHNOLOGY PERSPECTIVE, *supra* note 5.
9. *Id.*
10. *Id.*
11. IPCC, Press Release, Summary for Policymakers of IPCC Special Report on Global Warming of 1.5°C (Oct. 8, 2018).

will be locked in within 10 to 15 years—mostly emitted, too late to mitigate. The potential to reduce building operating emissions through design will be locked in as well. In other words, we have 10 to 15 years to get it right.[12]

While conserving energy for future generations is a primary tenet of sustainability, *thwarting global warming* is not a function of the amount of energy we consume, *but rather the release of its carbon content*. Although the efficiency of a building's design and construction predetermines its net energy consumption, the carbon intensity of its energy sources play the primary role in global warming. The carbon content of the energy consumed to produce the materials, for the construction process, and for ongoing building operations trump energy consumption. If *carbon-free* sources generated all of our energy—such as the sun, wind, water flow, gravity, temperature differential, or nuclear—we would not have these concerns. If carbon scrubbing technology and carbon sequestration were a large scale reality, perhaps they could neutralize the problem. Unfortunately, while carbon-induced global warming advances at an untenable rate, we remain far from realizing these solutions on a global scale: far from fossil fuel abandonment and mitigation technology reaching critical mass; and far from constructing sufficient carbon-free energy-generating systems, energy storage facilities, and the infrastructure necessary for distribution. Until this is remedied, the carbon embodied in our construction and the energy efficiency of our designs are the critical elements for emissions containment.

The Call for Action—Choice and Insight Matter

How significant is the building sector's contribution to GHGs? According to the IEA, the building sector represents the largest share of world's energy-related CO_2 emissions.[13] The built environment is the world's largest emitter of GHGs, emanating from both the construction and operation of our buildings and infrastructure.[14] Manufacturing building materials alone such as steel, cement, and glass produces 11% of the world's energy-related emissions. The need to decouple these emissions from economic growth is central to the IEA *Energy Technology Perspective* scenarios which involve reducing the carbon intensity of the world's primary energy supply and energy-efficiency measures. While noting the importance of long-lived core building

---

12. Computed from ENERGY TECHNOLOGY PERSPECTIVE, *supra* note 5 and IEA & UNEP, GLOBAL STATUS REPORT FOR BUILDINGS AND CONSTRUCTION: TOWARDS A ZERO-EMISSIONS, EFFICIENT, AND RESILIENT BUILDINGS AND CONSTRUCTION SECTOR 2017 (2017) projections [hereinafter GLOBAL STATUS REPORT 2017].
13. GLOBAL STATUS REPORT 2018, *supra* note 1, at 9.
14. ENERGY INFO. ADMIN., EMISSIONS OF GREENHOUSE GASES IN THE UNITED STATES 2009, at 22 (2011).

systems, the IEA emphasizes the primary importance of building designs, especially the building envelope, which "can last for decades or even centuries." And with the window of opportunity for change rapidly closing, swift and assertive measures are necessary given the long life of building sector assets.[15] Our call to action is perhaps best expressed in the IEA *Global Status Report* of 2017:

> An urgent focus on building envelope performance and design is needed including the policy levers and financing tools that enable affordable and sustainable building construction and renovations. Continuing the current pace of global activity would create a longstanding lock-in of buildings sector energy demand and subsequent emissions. Setting forth a "build it right from the start" approach (including building energy renovations) will help avoid unnecessary energy demand, as well as costly renovations later to improve underperforming buildings assets.[16]

Building it right from the start is essential, which calls for immediate action. This requires user-friendly carbon footprint disclosure, design finesse, and policies that account for "embodied" emissions.

---

15. *See* ENERGY TECHNOLOGY PERSPECTIVES 2017, *supra* note 5.
16. GLOBAL STATUS REPORT 2017, *supra* note 12.

Part Two
Thwarting Climate Change

Chapter Five

What We Are Up Against; What Must We Do

umanity's failure to confront the built environment's carbon emissions is not for lack of trying. During the last 20 years, scientists, engineers, architects, and designers have focused their attention on sustainable design as a means of energy *conservation*, and thereby the reduction of greenhouse gas (GHG) emissions to thwart the progression of climate change. As previously noted, more than 200,000 professionals have been trained and qualified by the U.S. Green Building Council for Leadership in Energy and Environmental Design (LEED®), and over 1.5 million residential units and 100,000 commercial and government projects have been registered or certified worldwide. The U.K. BREEAM® has registered or certified 2.5+ million buildings in 70 countries. Yet even with such training and certification oversight, more than 40%[1] of energy-related global carbon emissions still originate from the materials, construction, and operation of buildings. Though well intended, many of these sustainable design efforts were inappropriate, or misapplied applications of go-to antidotes under the broad umbrellas of "sustainability" and "green." Many resulted from oversights, unobserved footnote, or fine print qualifiers to the data and statistical conclusions. Many failed to focus on the most pressing need—to reduce carbon emissions now, not gradually over the next 10 to 15 years. Why, because the cumulative gains derived from operating efficiencies and zero-carbon energy over the next decade might be too insufficient to be effective soon enough. Yes, incremental efficiency gains in energy use are absolutely necessary, and incremental gains in the implementation of low-carbon and zero-carbon energy sources are essential. And yes, we must continue to safeguard our air, water, food chain, and environmental quality. Nonetheless, we must focus our attention and resources to the *immediate* reduction of GHG emis-

---

1. INTERNATIONAL ENERGY AGENCY (IEA), GLOBAL STATUS REPORT FOR BUILDINGS AND CONSTRUCTION: TOWARDS A ZERO-EMISSIONS, EFFICIENT, AND RESILIENT BUILDINGS AND CONSTRUCTION SECTOR 2018 (2018) [hereinafter GLOBAL STATUS REPORT 2018].

sions in order to buy time for cumulative operating gains and a dominance of zero-carbon energy to take hold. The thrust of *Thwart Climate Change Now* is to target the engine that drives this failure, "embodied" emissions—the carbon footprint attributable to the design of our built environment and the physical nature of our dwellings and infrastructure—their layout schema and materials. With all the exposure, training, and certification, the formulation of these catalytic elements is still taken for granted, is still a matter-of-fact; yet their very composition generates much of our environmental poison, determining what is emitted *before* a new structure is occupied. Think of embodied emissions as a mushroom cloud of GHGs released during fabrication and construction, forever reflecting back the earth's heat. Accordingly, the efficiencies achievable through physical design are paramount.

Building materials alone contribute 28% of all *building*-related carbon dioxide (CO_2) emissions. Just a single component of the embodied carbon, yet they generate 11% of the world's total energy-related emissions.[2] The release of carbon emissions occurs on so many fronts, it is difficult to stay abreast of the numerous contributors to establish meaningful priorities. Some seem obvious, but others are subtle. Had we confronted our reliance on fossil fuels a decade or two earlier, the rate of warming would have been more manageable. Efforts to reduce operating carbon would have been less burdensome, and reducing embodied carbon would have been easier and more effective. Unfortunately, that window of opportunity has closed. By hanging our hopes on the gradual decarbonization of the global energy supply from 2030 through 2050, we will have subjugated the future to wishful thinking. We are up against a need for *immediate* action to target the low hanging fruit that seeds coincident emissions, the physical nature of our built environment: the buildings, their systems, appliances, and infrastructure; the spectrum of materials they cause to be manufactured. Policy and design control this, yet policymakers and designers ignore the inherent carbon, as though the carbon footprints were imaginary, or future efficiencies justify design irrespective of embodied emissions. It is futile to evaluate a structure, an appliance or even the equipment to harvest zero-carbon energy without noting the emissions from its fabrication, maintenance, removal, and replacement. These are the key vectors of concurrent atmospheric carbon; they have a large and lasting impact on global warming. Carbon-free energy may be the only viable ticket to a long-term sustainable future, but we must forestall the acceleration of carbon-induced warming until it is available in sufficient quantities worldwide.

---

2.   *Id.*

Multiple means already exist to preclude emission increases and enable their ultimate reduction. In that regard, using less operating energy; switching to low-carbon and no-carbon fuel sources; smarter material selection, material re-use, and material repurposing; and renovating are all important. *All should be employed.* Nonetheless, some means offer more efficacy and timeliness than others. At this stage in the progression of global warming, the thrust to reduce emissions must not be misaligned with other concurrent issues of global concern. Pollution of our air, water, and food supply, each existential in its own right urgently threatens sustainability, yet resolution of each problem area requires its own laser focus. Some elements will benefit from the solutions of others, but combining them within a broad category such as a "green" or "sustainable" solution threatens to draw attention from one storm cloud to another, deflecting action from carbon emissions, the specific target that must be speedily addressed. Bundling the quest to minimize carbon emissions under a broad green umbrella, a broad moniker for sustainability, or an economic program, renders the probability of achieving a *timely* success minimal. Time is of the essence.

Whether tall or small, designing a building to be environmentally sound is synonymous with built-in operating efficiencies and energy conservation. Operating efficiencies and energy conservation dictate the level of a building's recurring energy consumption and thereby its annual emissions throughout its useful life. The tactical use of natural light, shade, insulation, sealed interfaces, the sun and the earth's warmth, and solar energy are typical means. Yet how often are the carbon gases emitted while processing and fabricating building materials the standard for their selection, let alone the transportation emissions or those from a building's construction and eventual demise? How often do we consider the manufacturing footprint of the appliances and conditioning systems? In other words, how often does a developer or designer contend *seriously* with a building's "embodied" emissions—its "carbon footprint"? Depending on the design and construction methodology, embodied emissions worth decades or more of a building's operating emissions can be avoided. Not only are they predetermined, a good part are emitted years prior during raw material processing, well before the construction begins.

Reading Beyond the Headlines; Probing the Text

Understanding what claims and statistics mean is a formative obstacle to effective carbon-conscious design, such as interpreting the headlines, the pronouncements, and targets in terms meaningful to those capable of effecting change. Most articles on this subject are either misleading or beyond relatable comprehension. As such, we lack a coherent basis from which to apply the tenets of appropriate design, or to evaluate alternate materials for an intelligent design decision. The manufacturers, suppliers, and industry associations that sell or promote the products dispense most of the technical information from which an evaluation can be made. Impartial reports with in-depth energy consumption and emissions data are difficult to parse. Restricted to a few use sectors with misleading titles they provide broad characterizations with numerous qualifications and exclusions. Expeditious solutions rely on *comprehensible facts* that do not require reading above, below, and between the lines. *The first stage of deciphering the problem is to "parse the claims"—fallacious conclusions can be disconcerting.*

Tracking progress on carbon emissions is akin to monitoring the stock market on an hourly basis; a zigzag of ups and downs with an ever-changing array of explanations and recommendations. It is not a sound basis for intelligent action without careful study. The underlying data from which most summations originate stems from just two sources: the International Energy Agency (IEA) for global statistics and the Energy Information Administration (EIA) for U.S. statistics. GHG emissions are estimates derived from energy production and consumption sales statistics. This is fairly sound but there are caveats. Owing to the complex presentation of this material it is not broadly understood nor easily analyzed. Consequently, storylines frequently ignore the subtleties and qualifications that would reveal their true meaning. Many news articles and trade publications draw talking points directly from an IEA's or EIA report's preface or executive summary without examining the detailed presentation of the underlying data, without evaluating "cause and effect." Cherry-picked statistics, unintentional or otherwise, can easily distort reality and lead readers to erroneous conclusions, for example touting significant reductions from buildings and construction in the United States while emissions increase to dangerous levels worldwide.

The well-meaning message to the American Institute of Architects (AIA) Committee on the Environment noted in *Part One: Chapter One exemplifies* the problem, the congratulatory pat-on-the-back emailed to more than 10,000 architects, designers, builders, developers, and academics. Without careful scrutiny: "Our hard work is having a BIG impact" might provide a false sense of accomplishment. Although the assertion that 2017 U.S. building sector emissions were 20% below those in 2005, *despite nearly 30 billion square feet of new construction*, was true, the hard work of the architecture and construction industries had little to do with those reductions, which were from power sector decarbonization and energy-use efficiencies—the declining use of coal and light bulb innovations that led the way by intent. They also had little to do with the diminished energy demand resulting from mild weather and poor economic conditions that were beyond our control. The 20% reduction *did not result* from building design nor construction improvements as might have been construed.

Fortunately, emissions declined *in spite of* the design and construction methodologies employed during the ensuing 12 years. There was upbeat news behind the numbers, but it was not appropriately emphasized. The increased utilization of compact fluorescent lighting followed by the introduction of light-emitting diodes (LEDs) provided a notable reduction in the energy used for illumination, and thereby a reduction in carbon emissions—perhaps the most significant contribution to lowering operating emission since 2000. With the replacement of incandescent lamps far from complete worldwide, the opportunity for continued reductions from LED lighting is very real. Replacing coal-fired energy with natural gas, wind, and solar was good news as well, although methane released during natural gas production is still problematic. These strategies significantly impacted carbon emissions; most importantly they were immediate and will continue to grow. But one should not confuse those gains with the lack of verifiable progress made elsewhere in design and construction. For those who read further in that holiday greeting, ominous warnings appeared which were far from the holiday cheer: "[I]f buildings and infrastructure are designed and built to current standards, we will lock-in *emissions that will be with us for the foreseeable future.*" And quoting U.N. Secretary General António Guterres: "[W]e are still not doing enough, nor moving fast enough, to prevent irreversible and catastrophic climate disruption."[3] Those statements are also true. Nevertheless, the message ended on a positive note: "Let's welcome in the New Year with some good news, and a resolution to increase our momentum!" *Momentum?* "Our hard work is having a BIG impact" may sound comforting in troubled times, but it nurtures a fantasy about the nature of our achievements, one that is misleading and potentially counterproductive.

The Data Is Available; Why Is It Confusing?

The primary source for energy production and consumption data in the United States is the EIA, with statistics from as far back as 1949. Established

---

3. Brady Dennis & Chris Mooney, *"We Are in Trouble." Global Carbon Emissions Reached a Record High in 2018*, WASH. POST, Dec. 5, 208, https://www.washingtonpost.com/energy-environment/2018/12/05/we-are-trouble-global-carbon-emissions-reached-new-record-high/.

in 1977 in response to the oil market disruptions of 1973, global warming was not an issue and energy-related emissions were not a concern. Though emissions became a topic in the following decade with a growing interest in sustainability, the EIA's principle purpose was to maintain efficient energy markets and data analyses for policymaking, which continues to this day. Emissions tracking was added to their charter in 1992. Given its origins, EIA reports are structured to provide analyses of energy generation and consumption, *not to scrutinize building construction and operating emissions*. As such, with data specifically categorized to track energy production and distribution, it is difficult to parse the emissions that are ascribable specifically to buildings and construction—those embodied and from operations. Energy consumption data is categorized by broad "end-use" sectors: Residential, Commercial, Industrial, and Transportation. That's it—just four, plus a single "supply" sector, Power, to track the electricity generated and sold to those sectors. The end-use sectors purchase electricity from the Power Sector, but they purchase fossil fuels and biofuels for space heating, industrial heat processing, and transportation directly from the suppliers. Some users harvest their own solar or geothermal energy as well. Energy sources include fossil fuels, nuclear, and the renewables—hydroelectric, geothermal, wind, solar, and biomass. Although these tabulations are seemingly straightforward, the extent of the energy consumption attributable to the building sector as a whole is not inherently obvious, and the ensuing carbon emissions are even less clear. When the EIA refers to Buildings, it refers to the composite operation of the Residential and Commercial sectors. Unfortunately, this presents only part of the built environment's picture. This characterization ignores the operating energy of industrial buildings as well as the energy embodied in *all* buildings. Therefore it ignores the resulting embodied emissions—their carbon footprint past, present, or future. The emissions tabulated for the building sector are computed solely from the energy "consumed" in daily operations. Some of the carbon footprint can be found in the Industrial Sector's output, though it is extremely difficult to allocate or parse. This might surprise those who base their headlines on EIA Building Sector statistics.

Most *embodied* emissions reside in industrial sector manufacturing along with its operating emissions. This covers a broad range of carbon emissions from raw materials and fossil fuel extraction to powering production equipment. The list spans the entire economy from consumer products to agriculture, forestry, fishing and everything relating to the physical aspect of buildings and infrastructure, i.e., heavy equipment, construction materials, appliances, and furnishing. It even includes some, but not all, emissions due to material transportation. Every material component's carbon footprint involves transportation: moving material from mines to processors to manufacturers to suppliers and to construction sites, locally or halfway around the globe. Some of these emissions are included in the Industrial Sector and some are embedded in the fourth use sector, Transportation, which comprises all vehicles whose primary purpose is to transport people and/ or goods. It does not include equipment or vehicles like cranes, bulldozers, tractors, and forklifts used in construction; those emissions are tallied in the Industrial Sector, but not necessarily attributed to buildings. Nevertheless, whether a byproduct of industrial production, transportation, or construction vehicles, these emissions result from a building's design and construction; their tally should not be obscured by sector categorization. We need to know what they are to understand the true impact of our buildings and construction on climate change.

As global warming knows no borders, however, the EIA's U.S. statistics are only part of the picture. Global statistics provide the overall portrait of GHG emissions and perhaps our fate. Those statistics are provided by the IEA, which also emerged from oil supply security concerns in 1973. The IEA did not track GHG emissions until the late 1990s; their *World Energy Outlook* focused on carbon emissions every two years as of 1998. IEA sector characterizations are similar to the EIA but not the same. This in itself creates confusion when discussing building sector emissions. The IEA categorize building information with nomenclature such as Buildings, Buildings Sector, and the Buildings and Construction sector. Whether referring to EIA or IEA data, the problem lies not in the nomenclatures themselves but in how they are defined, what they include or exclude, and principally *how they are perceived*. Unfortunately, EIA and IEA summaries for building sector energy consumption and emissions are often perceived as their *total* contribution, yet the embodied components are either absent or lost in the weeds. Sector differentiations in IEA global statistical presentations are much better than the EIA's, and many of their reports focus specifically on the buildings sector; but they too miss the mark when it comes to "carbon footprint." As a result, absent careful study, media portrayals provide misleading storylines that are easily misunderstood. Conclusions derivable from one agency's building sector reports may not reflect those from the other; especially when seeking cause and effect among the multiple facets of building design, construction, and operation. Unfortunately, "cause and effect" are prerequisite to finding a solution.

Missing Biomass and the Selective Lens

Another note of caution concerns the basis for "emission" calculations, which are *derived from energy consumption* data. CO_2 emissions are calculated by a mathematical conversion predicated on the carbon content of an *estimated* mix of energy sources: from oil, coal, natural gas, nuclear, biomass, hydro, wind, and solar. Moreover, the primary data is footnoted with qualifying assumptions visible only to those who read fine print and glossaries. For example, emissions from the use of biomass energy are excluded from residential, commercial, industrial, and transportation sector tabulations and the power sector as well. This conforms to the International Panel on Climate Change's (IPCC's) 2006 *Guidelines for National Greenhouse Gas Inventories*. Mention of these footnoted qualifications is generally absent from the abstracts and summaries often quoted. Those so inclined can include biomass emissions by doing the math, the data can be found at the end of the EIA *Monthly Energy Review*'s "Environment" section, which tabulates U.S. biomass consumption and emissions separately.[4] Although biomass emissions are considered to be air pollutants, biomass energy is not considered a significant source of carbon emissions in many circles because it is not generated from "fossil fuels." Plant-based biomass is considered nearly carbon-neutral because, until is used for fuel, it sequesters the carbon extracted from the atmosphere during photosynthesis. Unfortunately, being nearly carbon-neutral also comes with many qualifiers. GHGs are released during planting. Tilling the earth exposes soil to the air, releasing the earth's sequestered carbon. Fertilizer production, the use of farm equipment, transportation fuel, and the emissions embodied in farm equipment and buildings are all carbon intensive as well. It is true that carbon sequestered in trees or perennial plants can remain locked in their root mass for extended periods, nonetheless, deforestation for fuel has a significantly negative impact by eliminating an important source of natural sequestration.

Processing biofuels consumes energy and emits GHGs whether ethanol, wood pellets, charcoal, or other plant or animal byproducts. Growing agricultural crops one year but returning their carbon to the atmosphere in the next immediately negates the benefit. Deforestation for fuel is even worse in that regard, emitting CO_2 to the atmosphere daily while depleting a depository of long-term sequestration—one that will take decades to replace even

---

4.   GLOBAL STATUS REPORT 2018, *supra* note 1. According to the IPCC's 2006 GUIDELINES FOR NATIONAL GREENHOUSE GAS INVENTORIES, carbon released through biomass combustion is excluded from reported energy-related emissions. To reflect the potential net emissions, the international convention for greenhouse gas inventories reports biomass emissions in the category agriculture, forestry, and other land use, usually based on estimates of net changes in carbon stocks over time.

if replanted. Unfortunately, the characterization of biomass fuels as "carbon neutral" helps justify government subsidies that incentivize increasing production, rather than encouraging more appropriate zero-carbon renewables. In 2019, 217 million metric tons of carbon emissions spewed forth from the consumption of biofuels by residential, commercial, and industrial sectors in the United States. This added 7% more CO_2 emissions from energy consumption than were indicated from these three end-use sectors. Even more concerning, biomass energy constitutes nearly 45% of all primary "renewable" energy consumption in the United States; nearly double hydroelectric, double wind, 5 times solar, and 24 times geothermal consumption.[5] In 2011, the residential sector consumed nearly three-quarters[6] of the world's biomass fuel owing to its use by developing countries. It would be significantly more sustainable to use agricultural land to grow crops to feed people and livestock rather than to produce biofuel, and to replace biomass fuel with *zero-carbon* solar, wind, hydroelectric, and geothermal sources. Yet relying on the "true" zero-carbon renewables—solar, wind, and hydro—remains a distant goal.

Sadly, while *end-use* statistics provide an incomplete picture of carbon emissions, their repetition in published studies renders them credible. Conclusions generally cite the same source material, papers, and summaries. Waters are muddied further by deductions derived from differing time frames that refer to an assortment of base years (2000, 2005, 2010, 2016, and 2017) with forecasts to 2030, 2040, 2050, and 2060, or the end of the century. Comparisons are difficult at best but more often useless. Industrial production might be included, biomass fuels not, onsite transportation sometimes and the like. The results may apply to a particular statistic over an arbitrary time period filtered by an array of definitions; summarized in abstracts, executive summaries, prefaces, forewords, and press releases. News reports and journal articles are often contradictory or impossible to decipher, making it hard to validate trends, and worse yet, "cause and effect." Ultimately, the media and trade associations often provide spin from statistics that suit, unknowingly biasing conclusions, while manufacturers and developers might bias their promotional material by intent. Too often the *upfront* consequences of embodied emissions are missing, as well as their *impact on the global warming timeline.* This is difficult enough to parse in scholarly searches, but for those who write the news, who influence design, procurement, and policy decisions, factual verification is a complex puzzle not likely to be solved. The

---

5. EIA, Monthly Energy Review (Mar. 2020).
6. *See* IEA, Energy Efficiency Indicators: Fundamentals on Statistics (2014) and United Nations, United Nations International Recommendations on Energy Statistics (2013).

complete picture is not apparent. Embodied emissions are buried in industrial, transportation, and biomass fuels; and the 40% of energy-related CO_2 emitted by building and construction likely exceeds 50. Thus, a forum exists that proffers misleading conclusions, clouding the path to appropriate solutions. This need not be so. Data is available for analysis by those who scrutinize the charts.

Given the education and certification of design professionals, the abundance of emissions data available, and the present focus on sustainable design, we *have* the means to extract important revelations to improve our buildings and their methodology of construction. After all, with 11% of energy-related emissions attributable to material choice alone, more than one-quarter of the 40% total, we already know where to start. But in order to determine a pecking order for the actions, we must clarify which components require tackling up front. Our focus on sustainability is not misplaced, just misapplied, failing to heed the cause and effect timeline. The first step is to parse the claims, the second is to face "what we are up against." We are able to make sense of this picture. One measure of this challenge is expressed by the "emissions gap," the difference between the Paris Agreement signatory pledges and the emissions levels allowed by the 2 degrees Celsius (°C) and 1.5°C scenarios as they relate to "current" emission levels. The U.N. Environment Programme (UNEP) refers to this as the "commitment gap": "[T]he commitments countries are making to reduce their emissions and the impact these commitments are likely to have on overall emissions reduction."[7]

2019 to 2020

"Even if the nations of the world live up to their current commitments, that will likely result in global warming of around 3°C by the end of the century."[8] This stern warning was contained in the foreword to UNEP's *Emissions Gap Report 2018*. Not as a pronouncement of doom, rather it issued a call for immediate action, continuing with "yes, it is still possible to bridge the emissions gap to keep global warming below 2°C . . . . Closing the emissions gap means upping our ambition." But only one year later, the 2019 report foreword was more dire:

> Each year for the last decade, the UN Environment Programme's Emissions Gap Report has compared where greenhouse gas emissions are headed, against where they should be to avoid the worst impacts of climate change. Each year,

7. U.N. Environment Programme, *10 Things to Know About the Emissions Gap 2019*, Nov. 26, 2019, at https://www.unep.org/news-and-stories/story/10-things-know-about-emissions-gap-2019.
8. *See* Joyce Msuya, *Foreword* to UNEP, Emissions Gap Report 2018 (Nov. 2018).

the report has found that the world is not doing enough. Emissions have only risen, hitting a new high of 55.3 gigatonnes of CO_2 equivalent in 2018. The UNEP Emissions Gap Report 2019 finds that even if all unconditional Nationally Determined Contributions (NDCs) under the Paris Agreement are implemented, we are still on course for a 3.2°C temperature rise.

Our collective failure to act strongly and early means that we must now implement deep and urgent cuts.[9]

Sadly, the *Emissions Gap Report 2020* foreword contains the same refrain for the third year running: "Overall, we are heading for a world that is 3.2°C warmer by the end of this century."

The 2019 and 2020 gap reports reaffirmed that *cumulative* emissions from 2018 through the end of the century must be contained within 1,200 billion metric tons to achieve the Paris Agreement's most liberal goal, the *Below 2.0°C* scenario. With total emission including land use exceeding 55 billion metric tons per year, they be must curtailed to a maximum near 40 billion metric tons *annually* by 2030.[10] Furthermore, the energy supply must be "decarbonized by 2060 with 98% of all generation from low-carbon sources."[11] Simply put, by 2030 we must drive this annual 15 billion metric ton emissions gap to zero, and ultimately, "carbon neutrality" will be essential to maintain global warming below 2.0°C through 2100, removing as much carbon from the atmosphere as we emit. Nevertheless, decarbonizing our energy by 2060 is a long way off; our first challenge is to close the emissions gap by 2030. And that serves only a 66% probability of capping the globe's temperature rise near 2.0°C. More favorable odds of 80 percent[12] requires trimming another 820 billion metric tons *cumulatively* through the end of the century. *This is the long-term challenge.* In the short term, which we can impact now, we must reduce global emissions by at least 2.7% *each year* from 2019 through 2030 to maintain that 66% chance of keeping the increase below 2.0°C.[13] That means eliminating 1.5 billion metric tons in 2020, and a little less each year to 1.1 billion metric tons in 2030.

Overall, the breakdown of the 55.3 gigatonnes (Gt) of GHGs emitted in 2018[14] indicates the "causes" we have to mitigate, 2019 should be similar.

14.8 Gt: Energy-related emissions from buildings and construction

---

9.   Inger Anderson, *Foreword* to UNEP, Emissions Gap Report 2019 (Nov. 2019).
10. UNEP, Emissions Gap Report 2019 (Nov. 2019) [hereinafter Emissions Gap Report 2019]; UNEP, Emissions Gap Report 2020 (Nov. 2020).
11. IEA, Energy Technology Perspectives (2017).
12. *Id.*
13. Emissions Gap Report 2019, *supra* note 10.
14. *Id.*

14.2 Gt: CO_2eq from methane, nitrous oxide (N_2O), fluorinated gases[15]

11.6 Gt: Other energy-related industrial emissions

8.6 Gt: Energy-related transportation emissions

3.5 Gt: Land use change emissions

2.6 Gt: Other

At 14.8 Gt, energy-related emissions from buildings and construction constituted 27% of *all* emissions—of the 55.3 Gt total, not just those stemming from energy. Consequently, their 27% burden-share of the 15 Gt emissions gap equates to 4 Gt. Therefore, to achieve the *Below 2.0°C* scenario's minimum goal, emissions from buildings and construction must be reduced by 2.7% annually, eliminating 0.3 to 0.4 billion metric tons per year from each prior year through 2030.

The second largest contributor, CO_2eq GHG "equivalents" coming from methane, N_2O, and fluorinated gases, are usually absent from built environment discussions which focus on energy-related emissions. But at 14.2 Gt, they are a significant part of the total picture. Methane alone was responsible for 9.7 billion metric tons in 2018. 3.2 billion were attributed to the production of coal, natural gas, and oil. Fortunately, most of that will be eliminated in the long term through decarbonization of the energy supply. Nevertheless, 0.3 billion metric tons of methane will remain as long as we continue to use biomass fuels. Enteric fermentation from animal digestion and the decay of waste in landfills, wastewater and from agricultural manure management generated 2.6 billion metric tons of methane CO_2 equivalents. N_2O attributable to agricultural fertilizers, accounted for 2.6 billion metric tons of CO_2eq, and fluorinated gases, which are frequently substituted for ozone-depleting gases, are responsible for 1.7 billion metric tons.[16]

As energy-related emissions in general produced 68% percent of the total, all of the IPCC scenarios rely on decarbonization of our energy supply in the long term. That includes tackling the 14.8 Gt from buildings and construction. Nonetheless, as we are already behind in the quest for decarbonization, significant reductions must be achieved in the short term from thoughtful

---

15. J.G.J. OLIVIER ET AL., TRENDS IN GLOBAL CO_2 AND TOTAL GREENHOUSE GAS EMISSIONS SUMMARY OF THE 2019 REPORT (PBL Netherlands Environmental Assessment Agency, The Hague 2019).

16. The preceding statistics were computed from analyses and data supplied by the IEA & UNEP, GLOBAL STATUS REPORT FOR BUILDINGS AND CONSTRUCTION: TOWARDS A ZERO-EMISSIONS, EFFICIENT, AND RESILIENT BUILDINGS AND CONSTRUCTION SECTOR 2019 (2019); IEA & UNEP, GLOBAL STATUS REPORT FOR BUILDINGS AND CONSTRUCTION: TOWARDS A ZERO-EMISSIONS, EFFICIENT, AND RESILIENT BUILDINGS AND CONSTRUCTION SECTOR 2020 (2020); TRENDS IN GLOBAL CO_2, *supra* note 15; Global Methane Initiative, *Global Methane Emissions and Mitigation Opportunities*, https://www.globalmethane.org/documents/gmi-mitigation-factsheet.pdf; and U.S. ENVIRONMENTAL PROTECTION AGENCY, GLOBAL ANTHROPOGENIC NON-CO_2 GREENHOUSE GASES: 1990- 2030 (revised Dec. 2012).

building design and material choice. It is too late and too risky to rely solely on the "potential" of operating efficiencies and decarbonization, that they will be accomplished soon enough on a broad enough scale. In 2019, building sector operating emissions increased to a new high once again,[17] and total global emissions climbed to their highest level yet, 59.1 Gt.[18]

The Zero-Carbon Lens Versus the Built Environment's DNA

Zero-carbon energy is the ultimate solution for the building, construction, and manufacturing sectors. It will virtually eliminate "operating" carbon emissions and reduce the embodied carbon of future buildings, including their environmental conditioning systems and appliances. End-stage embodied carbon in preexisting buildings will be minimized as well: the embodied carbon attributable to maintenance, deconstruction and the processing of materials for reuse or disposal. Most importantly, the energy-related emissions embodied in future materials diminish when manufactured with zero-carbon energy; only emissions from the chemical byproducts of materials processing will remain, such as from processing cement and steel.

Multiple technologies are already in use to provide zero-carbon energy worldwide, harvesting and generating emissions-free energy; harvesting, the sun's energy with photovoltaic solar cells to output electricity, or with solar collectors or mirrors that heat fluids to create hot water or steam for electric generation. We harvest wind and water flow to mechanically power electricity generation. Nuclear energy rounds out the top four, while other means such as tapping the earth's geothermal heat, water current, and deep lake temperature differentials may gain traction as well. All are free of carbon emissions with the *notable* exception of *emissions embodied in their generating equipment, operating facilities and transmission infrastructure.* They all have embodied carbon, none of which is minimal and cannot be ignored. According to the IEA *Renewables Information Overview 2019*, "zero-carbon" energy sources produced nine percent of the world's total primary energy supply (TPES). In the long term, as this 9% share grows to 90% of the world's energy supply, the carbon embodied in future equipment, facilities, and infrastructure—manufactured, transported, and constructed with carbon-free energy—will become minimal too. Reaching this goal is a matter of financial investment, equipment availability, technological advances

---

17. GLOBAL STATUS REPORT 2020, *supra* note 16.
18. *Id.*

in energy storage, and the installation of sufficient storage and transmission infrastructure. And all of this requires time and "political will."

But once again we must choose our words carefully. The current focus is on increasing the use of "renewables." As with emissions statistics generally, renewables reports also can mislead or be difficult to decipher as they commonly refer to annual *growth*, rather than their total share of the energy supply. One must not misconstrue the percentage of "new" or "increased" capacity or generation, for the "total" capacity or generation actually installed. One must be careful not to understand "electricity" generation as our *entire* energy supply, nor to believe *all* "renewable" energy is "zero-carbon." The *Perspective on the Global Renewable Energy Transition*[19] reported that "[r]enewables accounted for 64 percent of new net electricity generation capacity in 2018," a staggeringly large number. But this refers to "new" capacity, not total generation. Renewables were estimated to reach 27% of total *electricity* generation worldwide by the end of 2019. This is a good accomplishment. According to the report, however, "[e]lectricity accounts for only around 17% of worldwide energy demand, so there is an urgent need to decarbonize heating, cooling and transport as well," as renewables "provide only 10% of the energy used for heating and cooling, and just over 3% of energy use for transport. Shares of renewables in these latter sectors are growing so slowly that renewable energy consumption is barely keeping up with global growth in energy demand." The *Renewables 2020 Global Status Report* concluded that "[d]espite the growing deployment of renewable energy around the world, the share of renewables in total final energy consumption (TFEC) has seen only a moderate increase."[20]

As of 2018, renewables accounted for roughly 13.5% of the world's primary energy supply and 12.3% of the U.S. primary energy supply.[21] This figure includes biomass fuels, which represented two-thirds of the global energy coming from renewables, and approximately 40% of the renewable energy in the United States.[22] Biomass emissions in the United States exceeded 300 million metric tons of CO_2 in 2020 for the 11th consecutive year, and were close to 7% of all U.S.[23] emissions from energy consumption, three-quarters from the residential, commercial, industrial, and electric power sectors. And biofuels represented 91% of the renewables consumed by road transportation

---

19. REN21 Secretariat, Perspectives on the Global Renewable Energy Transition, Takeaways From the REN21 Renewables 2019 Global Status Report (2019) (ISBN 978-3-9818911-7-1).
20. *Id.*
21. IEA, Renewables Information: Overview (2020); EIA, Monthly Energy Review (Jan. 2021) [hereinafter Monthly Energy Review 2021].
22. IEA, Renewables Information: Overview (2019) & Monthly Energy Review 2021, *supra* note 21.
23. United States biomass emissions sourced from EIA, Monthly Energy Review, tbl. 11.7 (Apr. 2021).

worldwide in 2017.[24] Biomass fuels are not carbon-free, though a significant improvement over fossil fuels, biomass emissions are far from negligible. Biofuels are helping to make this transition, but they too must be eliminated in the long term.

Given that eliminating high-carbon fuels and generating more zero-carbon energy are already top priorities on the world agenda, and that we are making gains, why can't we wait for energy decarbonization to cap global warming? While it is true that decarbonization will facilitate a large decrease in carbon emissions in some countries over the next decade, this will not happen on a global basis. Within the time frame currently required to close the emissions gap, decarbonization of the world's energy supply alone will not be widespread enough to cap the increasing rate of emissions, nor effective enough to slow the current rate of global warming. On the scale of the decarbonization required, this first step toward achieving "carbon neutrality" remains distant. Some believe such implementation might reach critical mass near 2050, while others believe it will be later. In all cases it will take a lot longer than a single decade to be completely resolved. Sadly, after three years of stabilization through 2016, with global emissions reaching record highs in 2017 and 2018, the portion due to energy-related emissions also reached record highs. Even if all of the commitments submitted by the signatories to the Paris Agreement were implemented by 2020, energy-related CO_2 emissions would still require a significant annual reduction by 2030, which must be maintained through the rest of the century. Seventy-eight percent of all emissions are produced by the G20 Group members[25] and there is no sign they will decline any time soon. Gap reports for 2018 and 2019 indicate that absent a rapid increase in action within the next few years, these emission levels *will not* reach their peak and stabilize by 2030 as desired. The "scale and pace of current mitigation action remains insufficient" and "current policies of G20 Members collectively fall short of achieving the unconditional [nationally determined contribution] commitments to the Paris Agreement.[26] As such, there is risk of long-term lock-in due to inertia, as well as a dependency on long-lived "capital stocks" and "committed emissions resulting from existing infrastructures."[27] This is the handwriting on the wall.

---

24. REN21 SECRETARIAT, RENEWABLES 2020 GLOBAL STATUS REPORT (2020).
25. Argentina, Australia, Brazil, Canada, China, France, Germany, India, Indonesia, Italy, Japan, Mexico, Russia, Saudi Arabia, South Africa, South Korea, Turkey, the United Kingdom, the United States, and the European Union.
26. GLOBAL STATUS REPORT 2018, *supra* note 1; EMISSIONS GAP REPORT 2019, *supra* note 10.
27. UNEP, EMISSIONS GAP REPORT 2018, *supra* note 8, 3.5.3.

We will continue to see improvements in the reduction of operating carbon emissions by municipalities around the world having access to low-carbon or carbon-free energy—especially wherever coal is eliminated in favor of natural gas, or more favorably by solar, wind, hydro, geothermal, or nuclear power. Nonetheless, whatever the rate, such implementation is unlikely to advance sufficiently to halt the progression of global warming within the next 10 to 15 years, let alone reverse it. Although each new announcement of a coal plant closed, a wind farm brought online, or a requirement for new homes to include solar panels provokes a sigh of relief, we must re-focus our lens to capture the global perspective. Global warming knows no boundaries. It is easy to become complacent, fooled by early gains associated with a specific nation or community whether it be the United States, China, the European Union, or otherwise, but the United States and most other countries are a far cry from the handful of those making major gains. It is easy to misjudge progress, fooled by those touting double or triple growth of a carbon-free source based on minimal prior use. Success or low emissions in one region, or conversion to low-carbon or zero-carbon fuel, does not necessarily indicate such gains for the entire global community.

We can no longer take solace from reports of individual strides forward nor can we wait for zero-carbon energy to become the world's predominant supply. Nor can we wait for global commercialization of technology to scrub CO_2 from the atmosphere. The urgency demands immediate reductions. Reducing operating energy expenditures, increasing the use of renewables, purchasing carbon credits, and the like are important steps in the overall scheme, but they do not justify the needless release of GHGs traceable to inefficient building design and construction, or unnecessary carbon embodied in the infrastructure we erect. These problems are more prevalent than one might expect, encompassing even the most rudimentary forms of construction. Constructing a simple 320-square-foot cinder-block shelter (30m^2) with a corrugated-metal roof[28] emits nearly 25 metric tons of emissions, just from manufacturing the cinder-block, mortar, and corrugated metal. Ten such structures would require 10 years of growth from 6,000 newly planted trees to absorb and sequester their embodied emissions.[29]

Slowing the buildup of atmospheric carbon in the coming decade is our most pressing problem. Tackling the immediate impact of embodied emissions, the near one-third of the built environment's 40% share, is an

---

28.   16 x 20 feet with 10-foot walls and a corrugated metal peaked-roof (30m^2, 5 x 6m with 3m walls).
29.   Calculated from "[r]educing 1 MMT of CO_2 emissions is equivalent to: 26,000,000 tree seedlings grown for 10 years," California Air Resources Board, www.arb.ca.gov, AB 32.

expeditious path to an impactful solution. Forethought and sound design can minimize a carbon footprint by dint of careful material selection and structural methodology. At the same time, designing an energy-efficient facade with a site-responsive orientation can reduce operating emissions as well. A building's physical design and construction establishes the demand for heating, cooling, mechanical ventilation, and artificial lighting, all of which engender operating emissions. Thoughtful design can reduce the need, thereby minimizing energy consumption and the associated carbon emissions. The less mechanical conditioning required, the less capacity required, the smaller the system procured, and the lower the embodied carbon. And as some structural systems yield lower carbon footprints than others, the same is true for cladding materials as well. The time spent devoted to low-carbon design is our best means for reducing embodied carbon, our best means to create *building efficiencies inherent to the design*. Design is doubly impactful; it determines both embodied and operating emissions for better or worse. Thus we can chip away at the carbon gap building by building. Millions upon millions of buildings are constructed each year and old ones renovated or reconfigured.

The operating carbon emitted today during the production of cement, steel, aluminum, plastics, glass, and the like will be included in future construction's carbon footprint. When these materials are purchased, delivered, and used in construction, they will be tallied as embodied carbon, part of a long trail of emissions related to the material's composition and manufacturing commencing with the extraction and processing of its raw materials. This is an emissions trail tied to material choice. The notion of a single structure's embodied carbon does not adequately acknowledge the environmental damage already incurred through the supply chain prior to its materials delivery, prior to a product's manufacture or a building's construction. This emissions chain is triggered regularly, renewed by each order for a new project, renewed by each order for stock that anticipates market demand, months or years before they are ever noted in a singular building's carbon tally. Each bag of cement, concrete block, steel beam, or pallet of bricks ordered renews the cycle for the next batch of stock, emissions released for an entire lot. As noted in the *Global Status Report* for 2018, the carbon embodied in our built environment is "primarily based on material demand." And worse yet, these upfront contributions to atmospheric carbon can take a decade or more of future operating reductions just to compensate. Moreover, not only do embodied emissions nullify the value of future operating gains, they kickstart an acceleration of the rate of global warming. In simpler terms, any reduction in embodied carbon retards the rate of change of global warming at the onset.

So What Is the Solution?

With the singular exception of wood, no ideal building material exists. There are no magic bullets for architects, engineers, or designers who seek to alleviate these problems. Growing wood removes carbon from the atmosphere, which it sequesters unless it decays or burns. Used as a building material, the carbon in wood is sequestered for the life of the building. But there are limits to its availability and a limit to the building height it can sustain. There are no ideal sustainable design techniques or products that are universally applicable, and few as beneficial as their marketing purports. Many fail to account for their carbon footprint, or their maintenance, efficiency loss while ageing or their eventual abandonment. Many fail to consider their end-stage carbon footprint or mitigating environmental damage caused by their disposal. When one wades through the fine print or calculates the embodied carbon in a design, surprises emerge. Referring to his award winning AIA Green Project, Larry Strain, FAIA, noted:

> My own "a-ha" moment on this front was when my firm calculated all the embodied carbon emitted from building the Portola Valley Town Center. It's a very efficient project and has performed better than expected, but when we ran all the numbers we found that construction still emitted 1,000 tons of carbon—roughly the same as 10 years of operating emissions.[30]

He was referring to a building certified LEED® Platinum.

When it comes to the building blocks of architecture, the most effective actions revert to material selection and the elementary practices of "reduce, recycle, and re-use." Though deceptively simple in concept, success requires meticulous design with a heavy emphasis on "reduce," both in mass and inherent carbon footprint—without waiting for decarbonization of our energy supply. According to the IEA in 2019, nearly 800 billion square feet (ft^2) of floor space (77 billion square meters (m^2)) will be built from 2020 to 2030 and 200 billion ft^2 (20 billion m^2) of existing buildings will be renovated; and sadly, "the global energy sector is not on track for a low-carbon transition." "Despite efforts to reduce GHG emissions, the world's energy supply still is almost as carbon intensive as it was nearly two decades ago."[31] Over this 10-year period, embodied carbon would be responsible for approx-

---

30. AIA, *10 Steps to Reducing Embodied Carbon* (Mar. 29, 2017).
31. IEA, Perspectives for the Clean Energy Transition: The Critical Role of Buildings (Apr. 2019).

imately 60% of these new buildings' emissions, two to more than three times the operating emissions released in 6 of those 10 years. 70 billion m² of renovations will increase the embodied emissions even more,[32] and none of this includes those attributable to the interior finishings, furniture, and fixtures that make buildings complete.

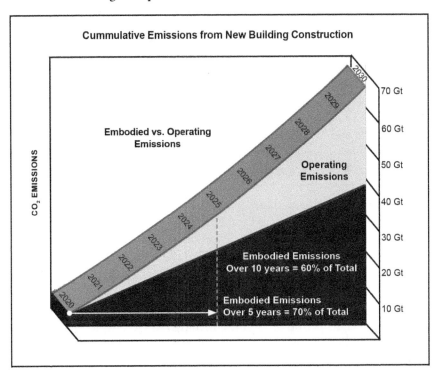

Accumulation of Embodied Emissions Versus Operating Emissions 2020-2030

Source: Graph generated from *IEA Global Status Report 2020* Building and Construction sector data for 2019 Emissions, prorated for 77 billion m² of new construction (*Perspectives for the Clean Energy Transition: The Critical Role of Buildings*, International Energy Agency (IEA), April 2019), with a Building sector emissions rate for a global building stock of 235 billion m² (*IEA Global Status Report 2017*).

---

32. Based on GLOBAL STATUS REPORT 2020 prorated for 77 billion m² of new construction (IEA, PERSPECTIVES FOR THE CLEAN ENERGY TRANSITION: THE CRITICAL ROLE OF BUILDINGS (Apr. 2019), with building sector emissions rate for a 235 billion m2 global building stock (IEA, GLOBAL STATUS REPORT 2017).

Carbon emissions incurred over the next decade cannot be reversed. We must reduce them through diligent design and construction. *The challenge is clear.*

What Hinders Low-Carbon Design?

We lack user-friendly tools to evaluate materials, to compare structures: to analyze their properties against their carbon emissions in terms relatable to building design decisions. Masonry, concrete, or steel; aluminum, vinyl, fiber cement, or glass? Or should it be wood? We also lack user-friendly tools to assess the value of energy-efficient appliances and building systems, their annual emission savings against those embodied and their useful lifespan. Current databases are awkward to use and lack useable content. Some materials are evaluated by weight, some by surface area, and others by volume—rarely relatable to products on the market without performing extensive calculations. When mitigating embodied carbon is the intent, evaluating alternative material choices can be difficult. Calculating a product's carbon footprint from an embodied carbon database can yield confusing and sometimes counterintuitive results at first glance. The culprit lies in how the source expresses carbon intensity. A material's carbon intensity—the primary indicator of carbon emissions—is derived from the mass of emissions released while processing and fabricating the product, from its raw materials to the factory's rear door.

Embodied emissions are typically expressed in kilograms or pounds (kg or lb) of emissions per kg or lb of the material: $kgCO_2/kg$ or $lbCO_2/lb$. As most construction materials and components are not designed in by the kilogram or pound, sometimes their emissions are expressed by surface area or by the piece: $kgCO_2/m^2$ or $lbCO_2/ft^2$ or per unit of use. But this too can be problematic: emissions per square meter or square foot depend on a component's thickness, when expressed for a single piece like a brick or a concrete block, they depend on its mass—such properties may vary from vendor to vendor. Comparing carbon intensities alone, without analyzing the property equivalencies, is another source for erroneous assumptions when applied to a construction project's carbon footprint. The requisite quantity for a specific use, i.e., the material mass, is the overriding factor in ranking emissions. For example, even though steel emissions are as much as 15 times more than concrete by mass, steel structures have lower embodied carbon than similar structures of concrete because steel structures have less mass. By material mass, concrete structures require significantly more concrete to carry a load than the amount of steel in an equivalent steel structure. Similarly, though aluminum and vinyl emit more embodied carbon per unit of mass than brick, clay, and ceramic—their thickness and mass are much less when used for cladding. As such, aluminum and vinyl *cladding* can be better than brick, clay, and ceramic depending upon the particular product's profile and composition.

Ranking materials by their emissions per unit of mass, from worst to best, yields surprising results:

Embodied Carbon—kgCO$_2$/kg[33]
Highest to Lowest Footprint by "Mass"

Aluminum
Polyvinyl Chloride (PVC) and Vinyl
Steel Sections
Glass for Glazing
Cement
Ceramic Tile & Cladding
Clay Tile
Common Brick
Concrete

But when accounting for the mass of their volume as typically employed in construction materials, the sequence of worst emitters is dramatically different. Concrete leads the list followed by steel. "Twice as much concrete is used in construction as all other building materials combined."[34]

Two heuristics that help identify *high* emission materials are based on the *temperature levels* required for manufacturing, and the *total mass* required for a project application. The *higher* the process temperature and the *higher* the mass used in construction *the more emissions* released. High temperature manufacturing places steel, cement, porcelain, ceramics, brick, aluminum, and glass high on the list when compared by unit of mass. Substantial energy is consumed while generating high temperatures, which in turn generates abundant carbon emissions. The second rule of thumb concerns the quantity of a material used; *the more material used the more the embodied carbon emitted*. This rule applies when selecting the thickness of tile or brick or even

33. Circular Ecology Ltd, *Inventory of Carbon and Energy (ICE)*, Database Version 3.0 Beta (Aug. 9, 2019): Aluminum: 6.6+ kgCO$_2$/kg with world average recycled content; PVC general: 3.1 kgCO$_2$/kg; Steel sections (beams, etc.) & plate: 1.6 - 2.5 kgCO$_2$/kg world average; Glass general: 1.4 kgCO$_2$/kg; Cement: 0.83 kgCO$_2$/kg UK average; Ceramic Tile & Cladding: 0.78 kgCO$_2$/kg; Clay Tile: 0.48 kgCO$_2$/kg; Common Brick: 0.21 kgCO$_2$/kg UK; Concrete general/average/high strength: 0.10/0.16/.19 kgCO$_2$/kg UK.
34. COLIN R. GAGG, CEMENT AND CONCRETE AS AN ENGINEERING MATERIAL: AN HISTORIC APPRAISAL AND CASE STUDY ANALYSIS (Elsevier Ltd. 2014).

glass; *the thicker the specific material or product the more the embodied carbon.* PVC and vinyl sit high on the list due to their chemistry. But many building products are a composite of materials, some a laminate of material layers such as many vinyl cladding products. What is the footprint of multi-layered vinyl cladding? That depends on the thickness of the vinyl. General guidelines can facilitate general comparisons, but cannot quantify the actual emissions. Heuristic guidelines can be helpful, but for meaningful whole building analyses, data-based values are a must.

For most architects and designers, existing product data are not sufficient to analyze construction schemes in their early stages of design conception. The design/build profession needs analysis tools that are tuned to commonplace trade specifications and ordering units, be they by weight, surface area, volume, linear length, or by the piece. They must have the ability to compare apples with bananas in the units they order as alternatives for the recipe. Current capabilities entail intensive research and spreadsheet work using a limited array of data sources with an inconsistent array of materials data. Better databases and computer programs dedicated to simplifying the carbon buildup *specific to building design* are crucial to intelligent selection, especially at the early stages of conception. Whether made available at no cost by an institution or a reasonable cost commercially, computer software of this nature could be developed within a year or two. Without this we remain handicapped, flummoxed by the carbon footprint of concrete versus steel, or of aluminum siding versus vinyl, or brick or glass; flummoxed with design choices for a proposed schematic. Differing compositions make this more complex as chemical treatment, finish, and other specifications for the same product may double emissions or more. Material choice is key to reducing carbon emissions, both embodied and operational, but product performance characteristics are key as well. Complexity, lack of clarity, and misinformation significantly hamper effective design.

Integrating sound sustainable methodologies with a physical aesthetic and the programmatic characteristics of the architecture is a complex task, especially after a building's schematic has been formatted. The lack of clarity regarding material and product performance render meaningful solutions a problem, and so do wishful fantasies and the trade's romance with the "symbols" of green design. This includes the inclusion of solar panels, green roofs, sun screens, double-skin facades, and the like, all valuable for specific site conditions—none universal for all. What is understood to be sustainable in general, often lacks sustainable value in application. Such reality is hard to decrypt even by architects, planners, and policymakers. Given the technical complexity and a general lack of transparency regarding performance requirements, this is not surprising. This holds particularly true for material selection in all facets of design and construction, from the specification of insulation to the methodology of a building's structure, and even to the colors of the brick from natural to grey or black. The choices made often encourage unworthy enthusiasm fed by misinformation, as well as certification or recognition for buildings that are environmentally unsound. Examples are lauded on a small scale while general construction on a large scale continues to ignore its contributions to atmospheric carbon, and false hope is proffered from the "potential" of undeveloped ideas. And sadly, we are up against marketing campaigns. The February 2020 issue of *Architect*, the official journal of the American Institute of Architects (AIA), contained a full-page ad courtesy of "Build With Strength," a coalition of the National Ready Mixed Concrete Association. The headline read:

**What building material absorbs carbon
for the entire lifespan of the building?**

CONCRETE IS THE ANSWER.

True, exposed concrete surfaces do absorb CO_2 slowly over time, as do some rocks and soil. But the decades required for meaningful absorption nowhere near compensate for the enormous emissions spewed daily by cement manufacturing. No, concrete is not the built environment's answer to eliminating the carbon problem, though some architects reading that AIA journal ad might be influenced to believe so.

Our 30-year flirtation with sustainable design has been insufficient to alter the course of global warming. Educated design focused on the reality of *net-carbon* can turn that around; targeting the embodied carbon first emitted, while designing for operating efficiencies at the same time. Carbon emissions attributable to each new building built, and each retrofit or renovation, can be reduced by design—both the immediate and the long term.

Chapter 6

The Role of Design

*Today's reduction is far more beneficial
than its future equivalent.*

The world's "urban" population is estimated to grow by nearly 1 billion city dwellers from 2018 through 2030, and 2.5 billion by 2050.[1] By 2050 we will have built more new urban areas and infrastructure to accommodate population growth than currently exist.[2] In a sobering report of April 2019, the International Energy Agency (IEA) concluded that delaying high-performance building construction and renovation for another 10 years would result in an additional 2 gigatonnes (Gt) of carbon emissions each year through 2050.[3] The IEA noted that "the number of new, high-efficiency buildings being constructed needs to increase more than 25-fold by 2030, with deep energy renovations of existing stock also needing to more than double within the coming decade." But once again, by focusing solely on operating energy consumption, they ignore the massive impact of the construction's *embodied emissions*. The year 2050 is a long way off, but 2030 is just down the road. How do we chip away at the built environment's emissions gap within the next decade while gradually decarbonizing the world's energy supply over the next 30 years? Three means are available: (1) *renovate* the current building stock for energy-efficient operation; (2) design *new* buildings to require *less operating energy*; (3) construct new buildings, infrastructure, and renovations with less embodied carbon. All three approaches can reduce the carbon footprint of the built environment. All of these approaches entail material choices and construction methodologies. And all will influence the

---

1.   United Nations, Dep't of Economic & Social Affairs, Population Division, World Urbanization Prospects: The 2018 Revision 1, 9 (ST/ESA/SER.A/420) (2019).
2.   Intergovernmental Panel on Climate Change (IPCC), *AR5 Climate Change 2014: Mitigation of Climate Change* 978, ch. 12 (2014) [hereinafter *AR5 Climate Change 2014*].
3.   International Energy Agency, Perspectives for the Clean Energy Transition 2 & 13 (Apr. 2019) [hereinafter Perspectives for the Clean Energy Transition].

net accumulation of carbon emissions for decades to come. Both "operating" and "embodied" emissions can be reduced through design.

Once material and construction emissions enter the atmosphere, long before a building's habitation, they no longer present an opportunity for mitigation. Energy efficiencies related to a structure's physical envelope, its environmental interfaces, are forever determined as well, locked-in once built. Both have everything to do with design. For this reason, the design and construction phases of what we build are uniquely impactful. Structure by structure, the design of our built environment determines its impact on global warming. We are fortunate in that regard as the next project designed affords an immediate opportunity to cut tons of carbon dioxide (CO_2) emissions. Moving forward, project by project, building by building around the globe, we can cut hundreds to thousands of tons of emissions per "blueprint" through educated design. It starts with the "bill of materials": the raw materials, parts, sub-components, and assemblies. We can reduce the demand for emissions-intensive materials upfront by scrutinizing material selection and alternative modes of construction. At the same time, we can design a building's envelope to reduce interior operating demands as well. With the aid of design, with means that already exist, we can help stabilize the atmosphere's carbon concentration before global warming reaches a tipping point. It will be tragic if we fail to face this challenge now as it is simply a matter of choice.

The CO_2-footprint of a building's structure and envelope, and energy efficiencies provided by its layout, orientation, and environmental interface, are the most important design elements that influence emissions attributable to the built environment. They all work together. The orientation of the envelope's walls influence the efficiency of the environmental interface, as do the energy transmissibility of its surfaces. The transmissibility of light, solar energy, heat, and air are functions of both design configuration and material choice. Material selection, providing the easiest opportunity to reduce embodied carbon emissions, is the key to our stopgap measures. In conjunction with sound design configuration, such determinations can curb annual operating-energy requirements as well and thereby the long-term cumulative impact of operating emissions. The energy-saving potential tied to envelope performance is very large, not only for high-performance building construction, but for the renovation of existing building envelopes. *Choices matter* not only for the materials selected and how they are configured, but for the abundance in which they are used. *Excess is a significant offender, yet the most elementary offense to tame.*

The Currency of Sustainable Design

The currency of sustainability permits only a minimal measure of emitted carbon. In the context of global warming, the less anthropogenic carbon we emit the better, the largest portion of which results from the design of our built environment: those embodied in its physicality and from environmental conditioning during operations. This includes the associated manufacturing, construction, and transportation emissions, factors often excluded from the analysis and thereby undermining the veracity of a sustainable design. Although embodied carbon occasionally receives lip-service during design discussions, it is generally taken as an inherent part of everything manufactured that we just have to live with. Thus, whether the focus is on conserving energy or reducing greenhouse gas (GHG) emissions, "design" solutions usually revert to obtaining operating efficiencies. And decarbonization, though not a primary function of design, consumes a lot of the attention regardless of its long-term timeline. We are blind to the carbon trail released during materials manufacturing and blind to the carbon trail we release during construction, both unleashing considerable atmospheric CO_2 to warm the earth's surface. To successfully thwart emissions growth over the next 10 years, those emission trails cannot be ignored. Our current age is responsible for an historic overuse of energy-intensive building materials with high-carbon footprints. This stems partly from choices made for ease of assembly and construction; partly for availability and cost savings; and partly for aesthetic design, for architecture as an art form, as a sculpture, or by decoration. Some of this is client driven and some is attributable to the designers; all of it lacks mindfulness to the impact, and none of it is currency for sustainable design.

Even with the incessant cry for sustainability and the media's acclaim for anything "green," there prevails a notable insensitivity to the carbon footprint in material choice; especially within the universe of architects and developers—much of it owing to insufficient information. Consequently, addressing the call for "green" by adding a few sustainability features to an unsustainable design has become commonplace. We see this with recent buildings on university campuses, residential buildings and even with architecture schools, many of which are Leadership in Energy and Environmental Design (LEED®)-certified and purport to be sustainably designed. The University of Miami School of Architecture's Thomas P. Murphy Design Studio, completed in 2018, provides an example. The Architect's Newspaper reported that the design was meant as a teaching tool to inspire future architects, "showcasing the basics of modern design, construction, and sustainability."[4] The orientation and design address the sun's position during the hottest months of the year with a 25-foot cantilevered roof overhang to shade the building's glass facade. The overhang curves downward at its southern tip with a wink toward sustainable design.

University of Miami School of Architecture, Thomas P. Murphy Design Studio.
Photograph ©Bill Caplan.

Unfortunately, the entire roof with its 5,000-square foot overhang consists of a 20,000-square foot slab of reinforced concrete, responsible for more than 100 metric tons of embodied emissions in addition to those from the studio's concrete walls.

4. *University of Miami School of Architecture Completes Its New Concrete Studio*, Architect's News-paper, Nov. 19, 2018.

University of Miami School of Architecture Thomas P. Murphy Design Studio Roof Overhang. Photographs ©Bill Caplan.

It is true that the mass of concrete architecture helps maintain interior temperatures under a hot sun, and that large hi-tech glass windows provide plenty of natural light while reflecting the sun's heat. Nonetheless, these sustainable characteristics do not justify this building's unnecessary contributions to global warming, particularly those emitted in bulk before the first student entered. A 25-foot cantilevered overhang in Florida's climate may well be a valid sustainability appendage—but not when constructed from concrete. One may question the use of concrete for the entire roof and some walls as well.

This is not unique to the University of Miami's School of Architecture. Many architecture schools around the world as well as institutions and commercial ventures are styled with a concrete aesthetic. In New York's financial district, 130 William Street, completed in 2020, is a 66-story, 800-foot-tall residential tower with a curtain-wall facade. The entire building is clad with arch shaped cast-concrete panels. The fact that concrete is a component of the building is not the issue, rather it is the overabundance.

130 William Street, New York City. Photographs ©Bill Caplan.

An abundance of concrete should be neither a teaching tool nor a design aesthetic to emulate. There are many lower-carbon material and design formats to configure and utilize; that is what architects are trained to envision. Yet, we address this currency of sustainable design as if it were an exercise in virtual reality. On the contrary, it is the "tangible" emission of CO_2 that must be contained as quickly as possible to maintain a favorable balance in the carbon cycle. We are bound by that reality and must confront it.

Concrete and steel are the world's two most widely deployed building materials. They were responsible for nearly 2Gt of the energy-related carbon emissions in 2017 from building construction alone.[5] Emissions from all uses spanning concrete garden furniture to steel vehicle frames were more than twice that.[6] Of the two, concrete is the more abundant emitter as it is used in significantly larger volume; not only because of its lower strength to weight ratio, but for its ubiquitous use in foundations, floors, walls, cladding, roofs and roof tiles, as well as roads, sidewalks, bridges, and infrastructure of all types. Concrete is the first step in upgrading from "flimsy" to "substantial" shelter: from bamboo, thatch, and corrugated-metal to structures from concrete block. Because a concrete structure requires significantly more concrete by weight than the weight of steel in a comparable steel structure, we are better off with a steel structure or a composite of the two.

Yet concrete itself is not the problem, it is the cement, the binder that secures the aggregate mix. Over four billion metric tons were produced in 2019 spewing two billion metric tons of emissions to our atmosphere. Although zero-carbon energy in the future will eliminate most materials' manufacturing emissions, some have carbon-based chemistry that adds to the problem. Cement is one of them, releasing carbon gases as a byproduct of its processing; not just from the energy that generates the heat, but from the material itself. More than half of cement's emissions are a chemical byproduct of calcination, breaking down limestone at 1,400 degrees Celsius (°C) to produce calcium oxide which directly releases CO_2. Only 40% of cement's carbon emissions result from burning fossil fuels in this high-temperature process; only those emissions can be eliminated by fuel-source decarbonization.[7] This is what renders cement particularly egregious. The carbon emissions released while processing cement amount to a ghastly 80% by weight of each bag. Depending on concrete's cement content, concrete emissions typically range from 10 to 20% by weight.[8] Several additional factors can significantly impact emissions, sometimes reversing one's expectations or heavily influencing the least-carbon choice. These include manufacturing location, fuel source mix, and recycled content. Where a raw material is processed or a product is manufactured dictates the energy supply's carbon content—think coal versus natural gas versus hydro. It also defines the point of shipment, the

5. PERSPECTIVES FOR THE CLEAN ENERGY TRANSITION, *supra* note 3.
6. IEA, *Tracking Industry* (2019), https://www.iea.org/reports/tracking-industry.
7. Johanna Lehne & Felix Preston, *Making Concrete Change, Innovation in Low-Carbon Cement and Concrete*, (Chatham House Report) (Royal Inst. of Int'l Affairs June 2018).
8. Circular Ecology Ltd., *Inventory of Carbon and Energy (ICE)* Database Version 3.0 Beta (Aug. 9, 2019).

emissions that will result from transportation. Pre-cast concrete panels fabricated in Dubai for a luxury high-rise in Miami[9] are more carbon-intensive than those cast in Florida. Both locations burn coal to process cement, but Dubai adds more than 9,000 nautical miles of delivery emissions.

Concrete, cement, and steel are necessary for construction, but we can reduce their use through more thoughtful structural design. Optimizing structural design can reduce the demand for concrete, cement, and steel by as much as 15 to 35% relative to a building's height.[10] Yet, the lack of attention to the carbon embodied in structural framing, or in infrastructure, often relates to prioritizing a specific design; perhaps a design firm's preferred framing methodology, redoing what they have done before. We are dealing here with the specification and relative mix of only three primary materials: cement, concrete, and steel—this is straightforward engineering and thoughtful execution of the aesthetic design, not "rocket science." Facades and interiors are more difficult challenges where layouts and material choices become more complex. A paucity of adequate information is the primary culprit, lacking an intelligible format and the means to evaluate carbon emissions with clarity. Choosing aluminum or vinyl or other plastics; brick or tiles of clay, ceramic or concrete; cladding walls, laying floors, and roofs. What surface percentage will be glazed—double or triple pane, which coating or film. Wallboards and studs and ceiling treatments. Numerous choices, products with vague compositions and minimal means to make comparisons. When it comes to embodied emissions we design in the dark. Careful material selection is very important as even "green" and "sustainable" products may carry unwelcomed carbon surprises. Many energy-saving building materials and design features contain greater embodied carbon than their conventional counterparts[11]; the emissions they might save over years of a building's operations might not exceed those embodied by their selection or excessive use. The selection and implementation of insulation materials provides an example. Although insulation is a primary tool to prevent the wasteful flow of energy, reducing the need for heating and cooling, it too carries a carbon footprint. Choosing low-carbon material is important as is minimizing the quantity used. Weighing the incremental gains in insulating value by adding thicknesses against the increase in embodied carbon, one might discover it is possible to have too much of a good thing. A buildup

---

9. Candace Taylor, *Zaha Hadid's Miami Tower Is an Architectural Feat. Is It Designed to Sell?*, WALL ST. J., Jan. 16, 2020 ("Ms. Hadid's design was complicated and expensive to execute, requiring nearly 5,000 pre-cast concrete panels to be shipped from Dubai.").
10. PERSPECTIVES FOR THE CLEAN ENERGY TRANSITION, *supra* note 3, at 55.
11. *AR5 Climate Change 2014*, *supra* note 2, 694, ch. 9.

of unnecessary layers of insulation is a popular practice in passive house and net-zero building designs. The Intergovernmental Panel on Climate Change's (IPCC's) *Climate Change 2014: Mitigation of Climate Change* noted that "[i]nsulation materials entail a wide range of embodied energy per unit volume, and the time required to pay back the energy cost of successive increments [of] insulation through heating energy savings increases as more insulation is added." This applies to some biomass products as well, when considering the emissions from feedstock.[12] Even biomass products are responsible for embodied carbon emissions. More insulation may make it easier to obtain passive house or net-zero energy conservation requirements, but this is not necessarily sound for reducing an accumulation of carbon emissions within the upcoming decade. At this crucial time while atmospheric carbon continues to accumulate at an alarming rate, we can no longer afford an unintentional splurge due to careless design. We need less carbon-intensive construction to kick-start the conservation. We need more attentive building design in our common forms of construction.

Notably, wood is absent from the foregoing discussion. As a construction material, wood sequesters carbon and is not a net emitter. Even though the energy spent to dress a pound of wood for construction emits approximately half a pound of CO_2, for milling etcetera and preparation, the wood continues to sequester the pound and a half it removed from the atmosphere during growth. In other words, wood building products have an effective negative "net" carbon footprint, storing a net of 1 lb of CO_2 for each 1 lb of dressed wood.[13] Burning wood for fuel , however, releases the pound and a half of sequestered carbon back to the atmosphere from which it came. Wood is good for construction if you can use it, but not for fuel. Although widely used to frame homes in North America, Oceania, and Africa,[14] nonresidential wood construction has been minimal and multi-story wood buildings have been limited to six stories throughout the United States. Fortunately, constructing multi-story wood buildings for other uses is gaining traction, and new building code categories pertaining to tall "mass timber" structures have been approved by the International Code Council. The 2021 International Building Code (IBC) will include codes formulated for 9- to 18-story mid- and high-rise mass timber buildings with heights to 270 feet

---

12. *Id.*
13. Circular Ecology Ltd, *Inventory of Carbon and Energy (ICE)* Database Version 3.0 Beta (Nov. 7, 2019): Fabricating Wood (Timber): 0.493 kgCO$_2$/kg; The wood Sequesters: -1.52 kgCO$_2$/kg; Net Embodied Carbon: -1.03 kgCO$_2$/kg.
14. IEA & UNEP, Global Status Report for Buildings and Construction: Towards a Zero-Emission, Efficient, and Resilient Buildings and Construction Sector 2018, at 45 (2018) [hereinafter Global Status Report 2018].

(82 meters (m)). Oregon and Washington State have already amended their building codes to include these mass timber categories, effective 2019 at a local municipality's discretion. Colorado has approved them for the city of Denver. As of this writing, Norway boasts the world's tallest timber building, an 18-story mixed-use building that houses a hotel, a restaurant, offices, apartments, and a public bath. Reaching 280 feet (85.4 m), it is built entirely in wood from glue-laminated timber (Glulam) columns, beams, and truss structures, and with cross-laminated timber (CLT) walls that bear secondary loads including elevator shafts and stairwells.[15] Engineered-laminated wood products such as CLT and Glulam have been around for a long time. Although they have gained acceptance in Europe, they have made little headway in the United States as major structural components, even for low-rise buildings, CLT was prohibited[16] for that purpose in New York City building codes through 2019. Several large-scale timber projects of up to six stories are expected in the United States for 2020-2021, but many more municipalities must adopt the IBC 2021 mass timber provisions to make a significant difference. The benefit of such buildings in the race to thwart global warming is extraordinary. Their low net CO_2 fabrication replaces the high emissions from manufacturing steel and cement; and while these components continue to sequester the carbon they store, new replacement trees planted in their stead will remove additional carbon from the atmosphere. The recent movement toward mass timber construction is very encouraging, and has the potential to alter the continuing spiral of embodied carbon—but only through widespread building with wood. France wants to set an example. On February 5, 2020, President Emmanuel Macron announced a plan that will require all state-financed public buildings to be constructed from at least 50% wood or bio-materials by 2022.[17] Macron's proposal was inspired by a recent Paris mandate: all structures eight stories or less built for their 2024 Summer Olympics must be of timber.

Small Buildings, Tall Buildings, All Buildings

Small Buildings

The sheer number of small buildings worldwide illuminates an opportunity to control the carbon destiny of this sector through the practice of informed design, a rather striking opportunity. Not only do residential dwellings comprise the largest share of small buildings, residences constitute the largest

15. Vollark, *Om Prosjektet*, https://vollark.no/portfolio_page/mjostarnet/.
16. Fred A. Bernstein, *Embodied Energy: A Primer for Architects*, Oculus, Fall 2019.
17. Charles Bremner, *Paris*, Times (U.K.), Feb. 6, 2020.

number of *all* buildings. The annual construction of new dwellings in the United States and the upgrade of inadequate shelters worldwide elucidate this reality, the two ends of the spectrum. At one end, approximately one million single-family homes are built in the United States annually with an average floor area in the vicinity of 2,500 square feet (230 m²).[18] At the other end, one and a half billion people live in temporary, minimal, or inadequate housing worldwide[19] which will be replaced by more substantial structures through the alleviation of poverty and forced displacement. The production of construction materials for each scenario will release carbon emissions annually that far outweigh years of potential operating emissions reductions in the future. Both scenarios offer opportunities to decrease this buildup of embodied carbon over the next 10 to 15 years through conscientious building design and material choice. Although a small building pales in carbon footprint relative to its urban high-rise counterpart, their relative numbers provide a different dynamic. This holds especially true when considering shelter for the billion-plus people who are impoverished or displaced. Initial upgrades in housing are typically from bamboo, wood, thatch, cardboard, fabric, and corrugated metal to cinder block and concrete, the worst material emitters on the list. To make matters worse, these upgraded replacements will be discarded and replaced as well, by multi-story structures on the march toward urbanization over the decades to come—more than doubling the emissions from this group, the largest of all. This first upgrade alone, though providing a minimally sized dwelling, will be responsible for releasing sizeable emissions. For example, more than a million and a half dwellings in Vietnam's 2009 housing census were classified as "simple" structures, constructed with supporting columns, roof, and walls that were all classified "flimsy"—constructed with materials such as mud, leaves, bamboo, thatch, or tar paper.[20] Replacing them all with minimal 320-square-foot dwellings (30m²) such as illustrated in the prior chapter—a mere four cinder-block walls and a corrugated-metal roof—would emit nearly 40 million metric tons of CO_2 to produce the building materials alone; a significant addition to the emissions gap. Building with brick would not fare much better, and clad-

---

18. U.S. Census Bureau & U.S. Dep't of Hous. & Urb. Dev., *New Residential Construction* (Aug. 16, 2019) and U.S. Census, Bureau, *Monthly New Residential Construction* (Sept. 2019).

19. Press Release, UN-HABITAT Strategic Plan 2020-2023, U.N. Habitat Assembly, May 2019 in Nairobi (Oct. 2, 2017), https://news.un.org/en/story/2017/10/567552-affordable-housing-key-development-and-social-equality-un-says-world-habitat.

20. CENTRAL POPULATION AND HOUSING CENSUS STEERING COMMITTEE, THE 2009 VIETNAM POPULATION AND HOUSING CENSUS: MAJOR FINDINGS (Hanoi, June 2010) (simple household dwellings, 1,666,071 (7.4%); temporary dwellings, 1,759,816 (7.8%); semi-permanent dwellings, 8,633, 005 (38.2%)).

ding with brick over cinder block would be far worse. This 40 million metric tons covers just the upgrade of four walls and a roof for the most inadequate housing in Vietnam. To provide minimally adequate shelter for the 1.6 billion people with such need, multiply that by 1,000. This is yet another hurdle for the emissions gap, created in good part by defaulting to cinder block and concrete rather than low-carbon or carbon-sequestering building materials such as bamboo and wood.

On the other end of the spectrum, specific to country and locale, single- and multi-family housing provides a different type of design challenge. In the United States, new buildings construction every year is equivalent to adding the entire building stock of New York City annually. Of the 1.2 million dwelling units built in the United States in 2018, 85% were framed with wood. Of the 840,000 single-family residences completed, 93% were framed with wood, and only 7% were framed with concrete.[21] Of the 345,000 multifamily units, primarily in low- and mid-rise buildings, 84% were wood framed. Only 3% were framed in steel, the balance with concrete or other framings. Although emissions reductions could be achieved if the 60,000 concrete structures built each year were framed with wood instead, even greater opportunity exists by addressing exterior wall materials. For example, only 5% of the single-family homes built in 2018 were also clad primarily in wood. Nearly 800,000 were clad with other materials, 26% vinyl siding, 25% stucco, 21% brick, 20% fiber cement, and 3% used other cladding products. Unfortunately, most of the materials manufactured for cladding either contain cement or use a cementitious material such as mortar for mounting. This includes stucco, fiber cement, thin clay-brick, and cut stone cladding products. Vinyl is the exception neither containing nor adhered with cement, but it too has a high-carbon footprint resulting from its carbon content. These cladding materials have an average manufacturing emissions in the vicinity of 7 kilogram $(kg)CO_2/m^2$, which varies widely as a function of the cladding's thickness, finish, and the volume of mortar used for mounting, if any. Vinyl siding seems to be the lowest, at less than half the average in its simplest composition and minimum thickness. Facing with full-size clay bricks on the other hand, which is not included in the above average, in the vicinity of 30 $kgCO_2/m^2$ can be four times worse than the mean and 10 times the vinyl.

21. U.S. Census Bureau, Highlights of Annual 2018 Characteristics of New Housing, https://www.census.gov/construction/chars/highlights.html (as of September 2019).

Unlike any of these materials, wood provides a benefit.[22] Wood shingles, shakes, or clapboards continue to sequester the carbon they have already removed from the atmosphere. If 20% of the houses in each non-wood cladding category were clad with wood, the net CO_2 emissions saved annually would near 1 million metric tons in the United States alone, solely from the cladding materials, solely from U.S. single-family homes. This does not include the backboard materials that support the cladding nor waterproofing membranes, mounting systems, or insulation, all of which can be selected to minimize a wall's composite carbon. The same can be said for window-frame, door, flooring, and roofing materials; choices among wood, plastics, metals, and ceramic, petroleum, or cement-based products. Such gains may seem small for each individual residence, especially in regard to a minimal shelter, but the quantities built and renovated multiply the impact; whether replacing bamboo and metal with cinder block, whether framing with wood, steel, or concrete, or whether constructing walls of brick, concrete, or low-carbon cladding. And as previously noted, each material purchase activates a re-supply chain, initiating a release of carbon emissions in the "now" time frame. Those emissions not only negate many years of operating emission savings over the very decade in which we must eliminate the emissions gap, the massive carbon they release immediately exacerbates the warming rate, increasing the likelihood that the gap will increase.

Single-family and small-multifamily dwellings are not the only structures comprising the *Small Buildings* category, let us not forget low-rise and mid-rise apartments, commercial and industrial buildings. Although significantly fewer than their one- to four-story residential counterparts, they are framed predominately with a combination of concrete and steel; often clad with fiber cement composites, concrete blocks, brick, clay tiles, aluminum, or steel—some claddings layered with a decorative or protective plastic-laminate finish. They too provide an opportunity to reduce embodied carbon through material selection and their structural methodology.

Tall Buildings

High-rise buildings are a more complicated affair. Material choice matters, but so does the choice of their construction scheme. The factors that make construction complex provide an opportunity to target embodied carbon. The Council on Tall Buildings and Urban Habitat classifies such buildings as "tall" (14 stories or more rising at least 50 meters/165 feet), to "super-

---

22. CO_2 emissions were computed from Circular Ecology Ltd., *Inventory of Carbon and Energy (ICE) Database* Version 3.0 Beta (Nov. 7, 2019).

tall" (above 300 meters/984 feet), and "mega-tall" for the few that reach 600 meters/1,968 feet. Their structural systems combine steel and reinforced concrete. "Reinforced concrete" structures are either cast in place over reinforcing steel bars or assembled from pre-cast components containing reinforcing steel bars. "Steel" structures are primarily steel even though their floor systems utilize concrete. "Composite" structures employ at least two of these systems as their structural members, such as steel columns encased in concrete, or a steel frame periphery surrounding a structural concrete core.[23] They all utilize concrete and steel, only their relative magnitude differs. The stability of a tall building requires a structural format specifically engineered to withstand the lateral wind loads and vertical gravitational loads that increase with each story, that increase with height. The most common formats are known as the "core-frame," "core-outrigger," "mega-brace," and "tube-in-tube."

"Core-frame" structures rely on a central core of walls interconnected to a building's rigid outer frame by its floor plates to restrain lateral movement at the higher floors. As a building approaches 60 stories, excessive bending of the frame's components will limit its rigidity unless additional structural augmentation is provided. The "core-outrigger" configuration is significantly better in that regard. The central core is surrounded by an outer perimeter of structural columns, all interconnected through its lateral outriggers—a network of high-strength horizontal and belt trusses that provide sideward support. "Mega-brace" structures avoid the need for the central core, utilizing diagonal braces external to a building's facade to resists both wind loads and the vertical force of gravity. Approaching 100 stories, a "tube-in-tube" construction is more suitable. With both an inner utility-core and an outer structural perimeter, the inner/outer structure resists lateral wind forces.

---

23. CTBUH Height Criteria for Measuring & Defining Tall Buildings, at skyscrapercenter.com and ctbuh.org.

Building Height Versus Carbon-Favorable Structure Methodology[24]

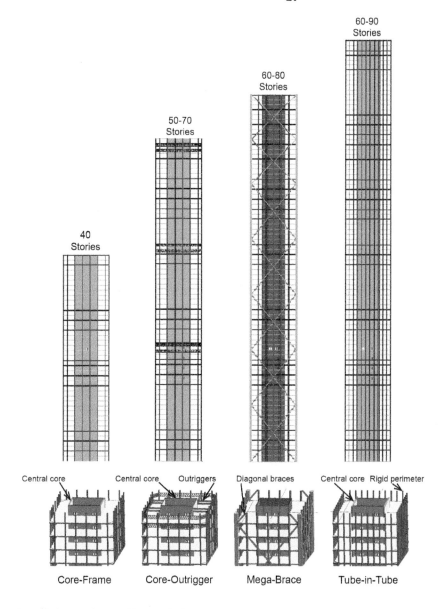

| Core-Frame | Core-Outrigger | Mega-Brace | Tube-in-Tube |

---

24. Author's illustrations based on drawings and descriptions in Vincent J.L. Gan et al., *A Comparative Analysis of Embodied Carbon in High-Rise Buildings Regarding Different Design Parameters*, 161 J. CLEANER PROD. 663-75 (2017).

Modeled in "[a] comparative analysis of embodied carbon in high-rise buildings regarding different design parameters,"[25] the impact of these four system's material emissions was evaluated, as well as the value of using recycled steel and altering the concrete's composition. The results are enlightening.

Unlike low-rise buildings where floor framing systems generally accrue the majority of the carbon footprint, the lateral-load structure accounts for roughly three-quarters of the embodied carbon in tall buildings, more than twice that of its floor system. In tall buildings, both the choice of materials to resist lateral loads and the overall mass will determine the magnitude of embodied carbon; but their raw material composition can influence the choice as well. Even though a "reinforced concrete" structure or a "composite" structure may have twice the overall weight of a "steel structure," building with *virgin* steel will produce 20 to 30% more embodied carbon.[26] Why? Because of the volume of steel necessary to create sufficient lateral force to stabilize the structure. Nevertheless, if the steel were processed from at least 80% *recycled* steel scrap, the steel structure would have a smaller carbon footprint than reinforced concrete or a composite. Specifying steel with a high recycled content makes an enormous difference. On the other hand, reducing the cement content in concrete can be a game changer as well. If the concrete in this application were able to be made with a cement substitute and maintain its full strength, a reinforced concrete structure would outshine a steel structure even if the steel were 100% recycled. Ground granulated blast furnace slag (GGBFS), a byproduct of producing iron and steel offers that potential. If low-carbon footprint substitutes for cement that could formulate high-strength concrete were widely available and verified, they would revolutionize the relationship between construction and embodied emissions. This would provide far reaching carbon emissions reductions for all uses of concrete, for all buildings and infrastructure, not just tall buildings. Nevertheless, in tall buildings, materials are only part of the equation, the relationship between embodied carbon and increasing height is related to the structural system. No singular form is best for all buildings; each system has a specific sweet region relating to embodied carbon versus height.

While the gross floor area of a tall building tends to increase linearly, the structure's embodied carbon tends to increase exponentially above a certain level. This relates to the increasing size of columns and core walls to support the increasing forces that occur with height. In the high-rise buildings com-

---

25. *Id.*
26. *Id.* §4.

parative analysis study, adding 20 stories to the 40-story model, to attain 60 stories, accumulated one-third more embodied carbon; the 20 floors from 60 to 80 stories added twice that. But 20 more to attain 100 stories was the worst yet, adding 4 times the incremental embodied carbon to obtain 60. On the other hand, as the height of a tall building is reduced toward that of a low-rise, the curve reverses and the carbon footprint per unit of floor area increases again, the base structure's embodied carbon represents a larger share. For example, graphing embodied carbon per unit of Gross Floor Area ($kgCO_2/m^2$) with increasing height, the "core-outrigger" format has its best carbon efficiency in the vicinity of 60 stories. The embodied carbon per square meter increases with more floors as explained above, but it also increases with fewer floors. Embodied carbon increases 4% per unit of floor area when reduced to 40 stories, and more below. A core-frame configuration is a better alternative at that height. Though 60 stories are core-outrigger's sweet spot, the study suggests a range from 50 to 70 is appropriate. A "mega-brace" is suggested for 60 to 80 stories and "tube-in-tube" for 60 to 90 stories.

Selecting the structural form best tuned to the intended height can make a significant difference, minimizing the carbon footprint per unit of floor area. After that, structural materials will determine the footprint's magnitude, either from "concrete," "composite," or "steel," as will the remaining list of building materials.

All Buildings

Regardless of building type or size, the less concrete used in framing, the lower the upfront emissions. Embodied emissions by floor area attributable to concrete and steel in "masonry-framed" buildings—framed with bricks, stones, or blocks—are about one-third less than those framed with reinforced concrete. Even better, the emissions from concrete and steel in steel-framed buildings are less than half of those framed with reinforced concrete.[27] Of these three choices, steel framing is the best by far. But none of those compare to the benefits of framing with wood. Unfortunately, outside of North America, Oceania, and Africa, many countries are heavily dependent on reinforced-concrete for residential framing, and in most countries outside of North America, concrete framing is most frequent for nonresidential structures as well.[28] Whether designing a small building, tall building, or infrastructure, we must confront its embodied carbon to minimize its net

---

27. *See* GLOBAL STATUS REPORT 2018, *supra* note 14.
28. *Id.*

emissions over the next 10 to 15 years. Neither decarbonization of the global fuel supply nor net gains from energy-efficient appliances are likely to provide the reductions required during that period. Yet the larger the reductions early on, the more effective they will thwart the increasing rate of global warming.

Chapter 7

Confronting Embodied Carbon Now

Current policies that address global warming generally promote energy-efficient appliances and energy conscientious consumption, though they *principally* rely on the future dominance of low- and zero-carbon energy. While the concentration of carbon dioxide (CO_2) in the atmosphere increases at a dangerous rate, and analysts worldwide tell us that the global warming rate is greater than anticipated, the worldwide adoption of sufficient carbon-free energy is far from probable by 2030 or 2035. Those policies are unlikely to bear fruit soon enough, nor cap the concentration of atmospheric carbon fast enough. According to National Oceanic and Atmospheric Administration (NOAA) data, the average annual concentration of atmospheric CO_2 globally has increased by 0.5% or more *annually* since 2010 at the very time we seek to reduce its warming impact. Increasing only 5.7% total in the 50 years from 1910 to 1960, the atmospheric concentration of CO_2 has increased more than 5% per decade since 2000.

Annual CO_2 Concentration (ppm)
Marine Surface—Globally Averaged[1]

Year	ppm	
1910	300.1	
1960	317.1	(5.7% over 50 yrs)
1980	338.8	(6.9% over 20 yrs)
1990	354.0	(4.5% over 10 yrs)
2000	368.8	(4.2% over 10 yrs)
2010	388.6	(5.4% over 10 yrs)
2020	414.0[2]	(6.6% over 10 yrs)

---

1. Calculated from data available Earth System Research Laboratories' Global Monitoring Laboratory of the National Oceanic and Atmospheric Administration, *Ed Dlugokencky and Pieter Tans, NOAA/ESRL,* www.esrl.noaa.gov/gmd/ccgg/trends. Data for 1910, 1960 and 1970 are "Ice Core Data Adjusted for Global Mean" from NASA Goddard Institute for Space Studies: Forcings in GISS Climate Mode, Well-Mixed Greenhouse Gases, Historical Data, https://data.giss.nasa.gov/modelforce/ghgases/Fig1A.ext.txt.
2. A record average monthly high of 417 parts per million (ppm) occurred in May 2020.

Global Warming Versus Atmospheric CO_2[3]

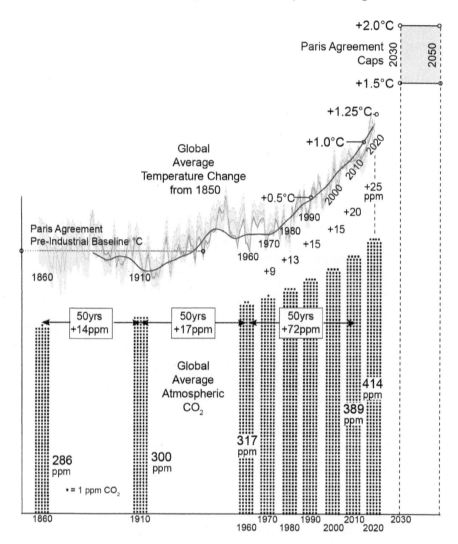

3.   The Global Temperature Change graphic is a composite of NASA/GISSTEMP v4 *Global Mean Estimates Based on Land and Ocean Data* (Lowess Smoothing Temperature Anomaly referenced to 1951-80), European Environment Agency (EEA) *Global Average Surface From PreIndustrial Reference* graphs from 1850 to 2020 generated by HadCRUT4, ERA5S, GISTEMP and NOAA Global Temp, and Berkeley Earth Org., *Global Average Temperature 1850-2019*, with ocean data adapted from UK Hadley Centre. Concentrations of Atmospheric CO_2 were calculated from data available from Earth System Research Laboratories' Global Monitoring Laboratory of the National Oceanic and Atmospheric Administration, Ed Dlugokencky & Pieter Tans, NOAA/ESRL, www.esrl.noaa.gov/gmd/ccgg/trends. Data for 1910, 1960 and 1970 are "Ice Core Data Adjusted for Global Mean" from

We have failed to address this accumulation of atmospheric CO_2 derived from materials processing and production, even though it may determine the soundness of environmental policy solutions that are slated for this decade. As embodied carbon creates a real-time surge in atmospheric carbon, the carbon footprint of those emissions cannot be excluded from policy and legislative considerations. The impact of embodied emissions should be an integral part of every policy assessment during the next 10 to 15 years, even those that do not target global warming. Additional means are needed to confront the rate of increase, to retard the current growth of carbon emissions in a short time frame. This includes re-assessing existing policies. It is too late to risk failure from unintended consequences that are foreseeable. It is too late to augment the carbon gap in this decade to obtain an accumulation of minimal "net" gains through the next.

The reductions required to address building sector emissions over the next 10 to 15 years can be achieved through policies and legislation aimed at material selection and building design. Such policies would weigh 10-year operational savings against the embodied emissions they would encourage, look for unnecessary embodied carbon in programs that encourage physical replacement, encourage energy efficiencies through building envelope design and orientation, and cap the embodied carbon in new construction as a function of floor area and use. While all building, construction, and manufacturing creates embodied carbon, intelligent choice and design can eliminate the unnecessary excess. Overlooking this reality has consequences, the worst of which will accelerate global warming.

Financial Incentives

Addressing climate change is big business, in good part related to acquiring "green building" credentials and certification. According to the International Finance Corporation (IFC), a World Bank Group Member, global investment in climate business solutions has exceeded $1 trillion annually and continues to rise: "About $388 billion of the $4.6 trillion spent on construction was invested in green buildings in 2015," noting that building envelopes make up the largest share of green buildings investments. The IFC foresaw market growth from government policies that can provide incentives to the private sector based on a "metrics-driven definition of what constitutes a green

---

NASA Goddard Institute for Space Studies, *Forcings in GISS Climate Mode, Well-Mixed Greenhouse Gases, Historical Data*, https://data.giss.nasa.gov/modelforce/ghgases/Fig1A.ext.txt.

building,[4]" which can raise public awareness to the benefits of green building ownership as well. They pointed to the use of low-interest loans, tax reductions, subsidies, expedited permitting, and green certifications as having a positive effect, reinforced by building codes, embedding energy-efficiency practices, and mandatory energy use benchmarking. Today, municipalities throughout the world subscribe to this approach, tying financial incentives to metrics-driven green building certification programs. The most prominent example in the United States is the use of U.S. Green Building Council's LEED® as the basis for certification.

For example, the city of Cincinnati, Ohio, incentivized developers with residential tax abatements for achieving LEED® certification, the higher the grade the better the benefit.[5] The maximum abatement permitted for building improvements was increased from $275,000 to $400,000 for achieving LEED® Silver, to $562,000 for Gold, and without limit for Platinum. The abatement term was increased from 10 to 15 years. On a national level, the U.S. General Services Administration (GSA) required all newly constructed or substantially renovated buildings to achieve at least a LEED® Gold rating, having stated in 2006 that LEED® "remains the most credible rating system available to meet GSA's needs." After its 2019 review of high-performance building systems, the GSA also included Green Building Initiative's Green Globes® program in its recommendations for new construction certification to the U.S. Department of Energy.

LEED®, BREEAM, and a few other certification programs have changed the face of the green building movement, both in the public's view and in architects' understanding of what can be achieved. By defining areas of design and construction through guidelines that enhance the built environment's contribution to human health and welfare, these programs help light the way toward environmentally responsible design. And as aptly stated in a 2003 White Paper by the editors of *Building Design & Construction*, a "LEED rating imbues projects with the equivalent of the Good Housekeeping seal of approval or a favorable review in Consumer Reports," "a branded metric that establishes a means of comparison in the real estate marketplace."[6] LEED® certification is dedicated to the phrase its acronym represents, Leadership in Energy and Environmental Design, leading the way toward energy-efficient and environmentally sound buildings. LEED® and other such programs provide well-suited general criteria for local, state, and national development

---

4. CREATING MARKETS FOR CLIMATE BUSINESS: AN IFC CLIMATE INVESTMENT OPPORTUNITIES REPORT (IFC 2017).

5. City of Cincinnati, Ohio, Ordinance No. 502-2012 (2013).

6. Building Design & Construction, White Paper on Sustainability (Nov. 2003).

projects, although some critics fault LEED® for the relative weight of its easy-to-achieve points. Out of the 110 points available for certification as of this writing, a mere 40 points are required to achieve the basic LEED® certification, almost half of which may have nothing to do with a building's design or its sustainability.

LEED® certification points may be accumulated for a site's proximity to transit, retail stores, restaurants, and housing; for locating in an economically disadvantaged community; providing affordable housing; or locating on previously developed land. Points are available for providing bicycle storage, for providing no more than the "minimum" allowable on-site parking, for selling parking spaces rather than including them in a lease, or for "not" providing off-street parking. Points are also available for purchasing "carbon offsets" and providing financial support to a conservation land trust or an accredited conservation organization within the region. Although 16 points are available for location and transportation, only 4 points are available for whole-buildings life-cycle assessment and 6 points for building product disclosures and optimization, none of which are mandatory. There is even a gratuitous point for including a LEED®-accredited professional on your LEED® project team. All of these are admirable and well worthwhile, but as a whole they fail to ensure the significant reduction of a candidate building's carbon emissions, nor require the containment of a building's embodied carbon. That does not mean that LEED® buildings are high in carbon intensity, it only means that LEED® as written is not an indicator of low-carbon design. *The most glaring fault of programs of this ilk is that they permit the occurrence of negative factors along with the positive without penalty.* The point system does not include a net score. A building can be certified by accumulating the required points from a variety of categories despite having unworthy design elements that actually negate its overall environmental value. From the 110 points possible, 40 points achieve LEED® Certified, 50 points achieve LEED® Silver, 60 points for Gold, and 80 points for Platinum, even if a building is constructed with an overabundance of concrete, or its solar panels are blanketed with shade, or the building systems need constant repair and replacement. One can earn points for procuring materials locally, but there are no subtractions for construction materials shipped from thousands of miles away. Nor are points lost for the gratuitous overuse of materials whether for lack of conscientious design or over-endowed aesthetics. What if the roof-mounted solar panels fail to generate enough energy to justify their embodied carbon footprint? Programs such as these as currently constituted are not suited to contain carbon emissions in a near-term time frame. They are not suited as a standard to thwart global warming.

When it comes to new construction, every little contribution to sustainable design helps, but a building or segment of infrastructure is either sustainable or it is not. Either its components (materials, construction, and assembly) and the entire entity's operation and maintenance collectively maintain a low-carbon footprint and energy efficiency over their entire life cycle, or they do not. If any appreciable segment does not, sustainability has not been achieved. Consider a new building with under-insulated walls yet fitted with *high-efficiency* triple-pane windows; or one with an all-glass facade in an un-shaded solar-intense environment equipped with high-efficiency air conditioning. One could create a case that these high-efficiency components reduce the total energy expended; nonetheless, such features are merely bandaids on an energy wasting design. They fail to mitigate the underlying problems and lack long-term benefits. Like all fittings and appliances, high-efficiency components age and deteriorate, require maintenance, are subject to failure or misuse, and have a limited life. They become less efficient, less effective, and will require replacement. All of this adds to their carbon footprint. Yet, while the precepts of sustainable design are common knowledge to designers and engineers, buildings with such design faults are still eligible for "environmental certification." With these faults or the overuse of carbon-intensive building materials, they still may be LEED® or otherwise certified. On a global scale, the reduction of carbon emissions attributed to the "sustainable design" of architecture is minimal at best. More suitably, they flag opportunities missed that will impact global warming for generations to come. All is not well in the pursuit of sustainability, and thus, all is not well in the pursuit of reducing carbon emissions by design. None of this is guaranteed by a green certification label. *Partial sustainability is not sustainable.*

Fixing Green Certification Programs

LEED® and other green certification systems do not require minimizing net carbon emissions. Although LEED® and other programs have brought the quest for energy-efficient buildings to the forefront, carving a way to achieve many energy-related carbon reductions directly or otherwise, scarcely a word is addressed to the containment of embodied emissions in their certification criteria. As of this writing, there are no requirements regarding the reduction of embodied carbon. Emit as much embodied carbon as your programmatic design and aesthetic dictate. In other words, don't waste time thinking about it. These are not primers dedicated to containing a particular building's carbon emissions. That can be changed—and it should be. Given the U.S. Green Building Council's powerful influence on the entire green building industry, and LEED®'s status as a standard metric to qualify for financial incentives, rebates, low-interest loans, and tax abatements that catalyze billions of dollars in investments, its certification criteria should be updated to heavily weigh the mandatory reduction of both embodied carbon and operating emissions. The endorsement of any green certification system by a government body should be preconditioned on its use of criteria for whole building emissions. Certifying criteria should require strict building standards that tally all carbon emissions attributable to a building, both embodied and operating; and not simply by intent, but by demonstrating compliance. They should be weighted to emphasize carbon-related points without including points for indirect attributions, such as those related to a site's location, the purchase of carbon credits or providing affordable housing. Those might be incentivized in local policies, and can be contained in a certification addendum, but they should not provide an easy path to environmental certification. A building's certification should be based on the totality of its own impact on the environment from its physicality and operating efficacy, and from its design and operation. To address the concerns of climate change, the issues at hand must concern the timeline of net emissions.

Net Carbon Thinking

To reduce the true growth of our building-related emissions we must address the lack of heed to their embodied component. Failing to account for such emissions from the products we manufacture significantly underestimates the magnitude of the net emissions released from energy-efficient features, energy-efficient appliances, and even zero-carbon energy-generating systems; from their manufacture through their first decade of use. The large plume of embodied carbon emitted *before* utilization hastens the rate of global warming, short circuiting their ability to retard annual growth for an extended period of time, yet we remain blind to its impact on the emission timeline. The gains forecast might not *begin* to accumulate until *after* 8 to 10 years of utilization; until embodied emissions are exceeded by cumulative operating savings. As a result, unintended consequences arise from a broad range of otherwise sound environmental policies, because many are applied too broadly and fail to consider net emissions. When the environmental cost of embodied carbon is not included in a policy's consideration, actions encouraged by green initiatives may prove unsound at a time when containing embodied carbon is paramount. Many practices and policies need more thorough appraisal, even those specifically designed to promote net-zero emissions or environmental conservation. Though they are seemingly beneficial for universal application, they too may inadvertently conjure substantial emissions absent the oversight of careful guidelines and restrictions. Placing solar arrays on every rooftop, replacing old appliances with more efficient models, and replacing old buildings with new construction all reflect impressive policy goals, but can be misguided efforts without careful limitations. *The devil is in the details, which are often camouflaged by the very analyses that make them seem sound.* Confronting embodied carbon starts with a check on reality, examining the complete picture. We cannot rely on "green" labels and blind universal application when time is of the essence. Gains derived over the first decade of operation that are nullified by embodied emissions *are not gains.* Given that carbon-free solar energy is a top priority and the installation of solar panels are widely promoted, a reality check on small-scale solar arrays is a good place to look for the pitfalls in a universal promotion.

Realistic Evaluation of Incentives

Small-Scale Solar Arrays

Installing solar arrays worldwide is central to a carbon-free energy future. Solar farms tied to the power grid and rooftop solar arrays constitute the lungs of this plan, drawing energy from the sun's radiation as our lungs extract oxygen from the air. Backed by federal and local tax incentives and local rebates, the installation of small-scale residential and commercial systems has become a major industry with a network of suppliers, installers, and financiers. Nearly two million residential photovoltaic (PV) solar arrays had been installed in the United States by the end of 2018, mostly on rooftops. Annual installations have grown nearly 35% through 2018 since the U.S. Energy Policy Act of 2005 created a 30% tax credit up to $2,000 for qualified residential PV expenditures.[7] Some states provided purchase cost rebates as well. Extremely successful, these policies encouraged the generation of carbon-free solar energy from rooftops across the country, immediately eliminating emissions from the fossil fuel energy they replaced. In 2014, New York City initiated a 10-year plan to install 100 megawatts of solar power

---

7.   David Feldman & Robert Margolis, *Q42018/Q1 2019 Solar Industry Update* (NREL/PR-6A20-73992) (May 2019).

by 2025,[8] a target that was subsequently increased tenfold to install 1,000 megawatts by 2030.[9]

Emission-saving benefits for large commercial solar farms are well established, but are they for *all* residential installations during their first 10 years of operation? A solar array's ability to make up for its fabrication and installation emissions relies on many factors that vary from installation to installation. Whether a system is sufficiently worthy or perhaps detrimental while we urgently seek to reduce emissions, is a puzzle that varies from one rooftop to another and often a function of an occupant's length of stay. Unfortunately, policymakers fail to adequately weigh the upside and downside of small-scale solar installations, and the magnitude of their solar-electric generation varies considerably with location, orientation, and maintenance. Accordingly, net emissions eliminated by replacing fossil fuels varies significantly, as do the years of operation required to achieve a net gain. But their embodied carbon emissions are uniform regardless of location; they are the same for a rooftop in southern California as a rooftop in the states of New York, Texas, or Washington. As net emissions over the next 10 years are of major concern, an installation's latitude matters, as does its unimpeded access to the sun. Utility-scale systems and ground-based solar farms provide a more economical ratio of operating carbon savings versus embodied emissions, and they can be situated in optimal locations. Financial investment in large-scale operations often affords procurement of advanced solar-cell technology and high-performance efficiency. Engineered for installation optimization and convenient access for regular maintenance, they are *key* to achieving zero-carbon energy on a global scale. This is not so for the universal installation of solar panels on every roof.

Beyond the core values of being "renewable" energy and "carbon-free," the rational for small-scale solar installations is primarily economic: savings for consumers who install them and profits for suppliers, installers, and banks. The leading indicator is the cost recovery time, determined by a system's cost and the value of the energy it generates, as supplemented by tax credits, rebates, and other financial incentives. Where leasing is involved, lease rate versus projected monthly electric bill savings provide the motivation. And to assess the savings, PV Watts®, an easy-to-use online calculator is available free to all, provided by the National Renewable Energy Laboratory of the

---

8. New York City, Mayor's Office of Long-Term Planning and Sustainability, One City: Built to Last: Transforming New York City's Buildings for a Low-Carbon Future (Sept. 2014), http://www.nyc.gov/html/builttolast/assets/downloads/pdf/OneCity.pdf.

9. New York City, Mayor's Office of Sustainability, New York City's Roadmap to 80 x 50 (Sept. 2016).

U.S. Department of Energy (NREL). NREL was "established to advance renewable energy technologies as a commercially viable option."[10] For any given location, enter the solar panels' kilowatt (kW) rating, angle of tilt, orientation, and the average local electricity rate. PVWatts® estimates the average kilowatt hours (kWh) per year the system will generate and the annual electric cost savings expected. Factor the financing, tax credits, and rebates over the life of the system and the economic value can seem very persuasive; hence two million residential systems were installed in the United States alone with more on the way. Although many homeowners enjoy their lower monthly electric bills, others fail to realize net gains due to lack of solar panel maintenance, unaccounted-for shade, system downtimes, or early abandonment. While this would be a financial downside for the homeowner, it is even more problematic in the quest to thwart global warming. The "sustainability" value of solar power has nothing to do with its financial gain, but everything to do with its net carbon benefit versus time, in this case over the next decade. This too is not particularly transparent without careful investigation.

A common measure used to justify an installation's worthiness—its sustainability—is the system's energy payback time (EPBT). EPBT represents the number of years of solar generation required to offset the energy consumed for manufacture, which provides an indication of the years of use required to neutralize an installation's embodied emissions. Solar panels for residential use have commonly been advertised with an energy payback time in the vicinity of five to six years, some producers claiming three years or less. Both ranges seem like a "green" investment, lauded by governments and environmental groups worldwide, incentivized by rebates and tax credits and immediate reductions on electric bills. This is ideal for all. Solar electric generation relies on local solar irradiance, the amount of solar energy that strikes the panels at a particular site, expressed in kWh per square meter of surface area (kWh/m^2). The industry standard for solar panel comparison is typically 1,700 kWh/m^2 per year, considered a common irradiance for southern Europe and parts of the United States. Unfortunately, not every location receives 1,700 kWh/m^2 per year. Although southern regions in the United States receive sufficient irradiance to suggest a short EPBT, northern states from coast to coast have 10 to 25% lower solar exposure; payback times increase accordingly.[11] With the inclusion of financial incentives, a few extra years over the long term might not make a difference from an investment point of view, but they do in terms of the embedded carbon emissions they

---

10. *See* Feldman & Margolis, *supra* note 7.
11. EIA, ELECTRIC POWER ANNUAL tbl. 3.2 (2018), https://www.solarenergylocal.com/states/texas/dallas.

entail. When grappling with global warming in real time, EPBT predictions are a misleading metric, as are some of the illustrations used to promote small-scale residential solar systems, such as expression of life-cycle greenhouse gas (GHG) emissions in "grams" per kWh—a unit of mass more suitable to baking a cake than to the "metric tons" per typical rooftop system. Promoting the use of solar energy, NREL compares PV solar systems against coal-fired electricity. Only 40 "grams" of CO_2 equivalent per kWh of life-cycle emissions from crystalline-silicon solar cells compared to 1,000 grams for coal—comparing the best with the worst. As if all the energy replaced would come from the dirtiest fossil fuel of all.[12] Comparing life-cycle stages and operations emissions from each, they note that coal-fired power plants emit the vast majority during operation. For PV power plants, the majority is upstream in materials and module manufacturing. True, which means that every PV system commences with a large emissions footprint, one that requires neutralization before net benefits can accrue—a point that was not made. Some literature written to enlighten decision makers and investors reads like a "greenwash," masking what careful scrutiny might foretell. A more realistic look at residential and small-scale business installations reveals that neutralizing embodied emissions in the near-term requires an appropriate location and orientation. To make the investment of embodied emissions worthwhile also requires vigilant maintenance and many years of continuous use.

So, what do the published numbers mean, how was "40 grams" of emissions per kWh derived, does it truly represent a typical rooftop installation's worthiness? If a crystalline-silicon PV system generated direct current (DC) electricity for 30 years without downtime, exposed to an average solar irradiance of 1,700 kWh/m^2 per year, the answer is "yes." That assumes no downtime for the DC to alternating-current (AC) inverters as well, and that all of this electricity is "utilized" onsite or through distribution. Unfortunately, as these conditions are not typical, the claims are greatly overstated. They ignore the wide variety of installation locations, operational characteristics, component failures, and their consequent ramifications for the first 10 years of performance by failing to generate sufficient carbon-free energy to compensate for the emissions invested upfront—the embodied carbon. The claims also ignore another indicator of useful life, manufacturers' warranties. The average warranty for solar panels is only 25 years, not 30, and assumes that a continuous "decline" in electricity generation will occur over

---

12. National Renewable Energy Lab., *Life Cycle Greenhouse Gas Emissions From Solar Photovoltaics* (Nov. 2012) (NREL/ FS-6A20-56487).

that time. Warranties for the electrical components required are more typically 15 years or less, and the average U.S. residence changes ownership in 7 to 9 years, at which time many systems are abandoned or leases go unrenewed. In reality, whether a system were in use for 30 years, 25 years, 15 years, or less, with downtimes for maintenance, storm damage, repair and replacement, and perhaps early abandonment, its operation would rarely be continuous. Other operating losses come from less than ideal panel orientations, from unaccounted-for shade, especially in treed suburbs and dense urban areas, and from lower solar irradiance in northern climates. Rooftops with the wrong pitch or facing a direction receiving less sun, buildings partially shaded by trees or tall buildings, or vertical facade-mounted panels, significantly impact the outcome. When it comes to the payback of embodied carbon emissions, the "average" scenario has no value—the details matter, of which there are many.

A more appropriate metric would be a statement of each solar panel assembly's embodied carbon. The manufacture of a "6.4kW DC"-rated residential array, the median U.S. residential installation in 2018,[13] has emitted approximately 11 metric tons of carbon. It is that simple. If you divide 11 metric tons by the number of kWhs theoretically generated and consumed over 30 years under 1,700 kWh/m^2 per year of solar irradiance and ideal operating conditions, the outcome will indeed be the feeble sounding range of 40 grams[14] of carbon emissions per kWh. Perhaps to capture a more realistic understanding, we should determine the number of years necessary for each particular locale to generate enough carbon-free electricity to pay back that 11 metric tons of emissions before benefits actually accrue. Tagging the system with a "CO_2-footprint: 11 metric tons" label would certainly clarify the situation.

For example, consider Pittsburgh, Pennsylvania. A "6.4kW DC" PV panel rooftop array in Pittsburgh, aligned with the most advantageous orientation and tilt, would receive close to 1,700 kWh/m^2 per year of solar irradiance. According to NREL's PVWatts online program, it would generate approximately 8,000 kWh of AC electricity each year. Based on 2018 reports, the U.S. retail electric generation it would replace emitted an average of 0.174 kilograms (kg) of carbon emissions per kWh supplied to consumers.[15] Therefore this rooftop system would eliminate 1,400 kg of emissions per year, 1.4

13.   Galen Barose & Naim Darghough, Tracking the Sun: Pricing and Design Trends for Distributed Photovoltaic Systems in the United States (Lawrence Berkeley Nation Laboratory Oct. 2019).
14.   Computed with NREL PVWatts "standard" solar array efficiency of 15%, used by NREL through July 2020.
15.   Computed from the DOE/EIA-0035 (2019/11), EIA, Monthly Energy Review (Nov. 2019).

metric tons. At best, this ideal rooftop array would require eight years of uninterrupted operation to compensate for its 11 metric tons of embodied emissions—emitted before the first useful kW was generated. Unfortunately, over millions of installations, "best" is not the norm. With just a few hours of partial shadow daily, a few days covered in snow, or less than the optimal orientation, an 8-year payback quickly degenerates to 10 years or more. Systems sustaining damage or not maintained, with faulty components or abandoned by the second or third owners, may never see the gains. Southern areas such as Phoenix, Arizona, enjoy a better picture. But even there under optimal conditions, a residential "carbon" payback of less than five years is unlikely without the most advanced solar-cell technology, a more costly investment. With the 10- to 15-year time frame of our present concern, perhaps 5 years is tolerable, but certainly not 10.

Estimates indicate that in 2018, new residential solar systems in 22 northern U.S. states with less than 1,700 kWh/m^2 per year of solar irradiance, brought online 736,000 megawatt hours of electricity generation. This constitutes communities above 36 degrees north (°N) latitude in the East and 41°N in the West[16] and roughly equates to 90,000 small-scale residential arrays.[17] Their embodied carbon—emitted prior to startup—equals nearly one million metric tons of CO_2, resulting largely from government-sanctioned solar energy rebate and tax abatement programs.[18] Although rooftop systems in southern states exposed to more than 1,700 kWh/m^2 of solar radiation per year are a good investment, as are most commercial solar farms nationwide, most small-scale rooftops PV systems in the northern half of the United States, residential and commercial, are more apt to *increase* the carbon gap through their first 10 years of operation. Because the impact of embodied carbon is neither transparent nor obvious, even cities with very progressive environmental policies continue to make this mistake. New York City is one of them.

New York City's Local Laws 92 and 94, effective November 15, 2019, require that new buildings, new roofs resulting from enlargement of existing buildings, and buildings replacing existing roof decks, install a solar PV-generating system of at least 4kW capacity, or a green roof system, or a combination thereof. This is subject to exceptions such as for rooftop equipment,

---

16. Computed from EIA, *Electric Power Annual 2018* estimates for New England, Mid-Atlantic, East North Central, West North Central without Kansas, Oregon, and Washington State.

17. Estimated from the number of 6.4kWdc residential arrays required to generate 736,000 megawatt hours of power in this geographical sector.

18. TRACKING THE SUN, *supra* note 13: "Financial incentives provided through utility, state, and federal programs have been a driving force for the PV market in the United States."

fire access, designated uses, or if a 4kW system cannot fit. State and federal tax abatements for PV systems, as well as the opportunity for electric bill savings, provide an incentive for the solar panels rather than the green roof alternative. As in Pittsburgh, a rooftop system in New York City with an ideal orientation, absent shadows from other buildings and a coating of dust or soot will require nearly eight years of uninterrupted operation to offset the tons of embodied emissions. But the average system, unlikely to meet such "ideal" installation standards, will more likely require 10 to 15 years of uninterrupted service. Every 100 installations of this minimally sized PV system will add nearly 700 metric tons of embodied emissions. As the solar panels are made elsewhere, those emissions will not add to New York's tally and thus ease the way toward reaching the city's goal. Nonetheless, New York City's gain is a global loss, retarding efforts to reduce climate change now.

As the median size of residential installations in the United States has reached 6.4 kW DC, the embodied emissions per system are more likely to exceed 11 metric tons in New York and elsewhere, compounding the problem. Nearly 95,000 new installations estimated for Illinois over the next five years[19] would emit more than 1 million metric tons of embodied carbon, 11 metric tons each, before generating a single kW of electricity. With 10 or more years of solar generation required simply to neutralize those imprudently timed emissions, perhaps the installation of small-scale PV systems in the northern climates around the world should wait. When policies that mandate or incentivize such installations turn a blind eye to *factual* operating conditions, they produce unintended consequences. Some may argue that emitting 1 million metric tons of embodied carbon over five years in Illinois, or in a single year from the 90,000 northern U.S residential systems installed in 2018, pales in comparison to the emissions they would save in the future by replacing coal. But that is a false comparison. Coal accounts for only 10% of *residential* electricity consumption in the United States and worldwide[20] while all of these low performing installations will add CO_2 emissions to the atmosphere by day one. Furthermore, far more than one million metric tons of upfront emissions are at stake. Residential rooftop systems worldwide are expected to increase by nearly 50 million in the 5-year period through

19.   Press Release, Wood Mackenzie Power & Renewables and the Solar Energy Industries Association, United States Surpasses 2 Million Solar Installations, May 9, 2019, https://www.seia.org/news/united-states-surpasses-2-million-solar-installations.

20.   Calculated for 2018 from the IEA, *Global Energy and CO₂ 2019 Status Report: The IEA Electricity Information Overview* (2020) and the EIA, *Monthly Energy Review* (Mar. 2020).

2024.[21] Almost one-third[22] of those systems would be situated in population areas with a *mean* annual solar irradiance between 1,055 to 1,630 kWh/m^2 per year under ideal mounting conditions. About 150 million metric tons[23] of embodied carbon would be emitted before being brought online for residential use, requiring 8 to more than 12 years of continuous operation before providing a net emissions benefit.

The same principle applies to the replacement of otherwise sound appliances for the sole purpose of obtaining operating savings. The new model's embodied carbon cannot be ignored. New "energy-efficient" appliances consume less energy than old ones, but do their 10-year savings outweigh the emissions incurred during manufacturing? Despite the good intentions of many "replacement" policies, their initial emissions can actually move us further from the near-term goal, adding to the near-term gap at a time when we must close it with urgency. Assuming that the current influence of their embodied emissions will be compensated by operating efficiencies, or perhaps decarbonization 10 to 30 years hence, is a risky premise given the high level of CO_2 already in the atmosphere and its persistent rate of increase. When an energy savings or emissions reducing solution entails the substantial release of new emissions that require a decade just to nullify, the solution will exacerbate the problem from the start. With the exceptions of constructing infrastructure to connect carbon-free energy to the power grid, and manufacturing equipment to store carbon-free energy and the like, we cannot risk the impact of emissions embodied in what we manufacture and build matter-of-factly, especially while facing an existential situation in which time is of the essence. This also applies to policies having nothing to do with saving energy or reducing carbon emissions, but are meant to secure our collective welfare though other means. Low-flush toilets come to mind, a policy "no-brainer" that conserves water. Unlike the benefits from solar electric generation which vary considerably with location and orientation, there is a universal water-saving benefit from low-flush toilets wherever they are flushed. Nevertheless, when it comes to "replacing" old, less-efficient models, this too can exacerbate global warming.

---

21. IEA, Press Release Global Solar PV Market Set for Spectacular Growth Over Next 5 Years (Oct. 21, 2019).

22. Calculated from 2010 populations of 2.122 billion in 33 countries with less than 1635 kWh/m^2/yr mean irradiation weighted by population for most favorable fixed system tilt. Data sourced from C. Breyer & J. Schmid, Population Density and Area Weighted Solar Irradiation: Global Overview on Solar Resource Conditions for Fixed Tilted, 1-Axis and 2-Axes Pv Systems, https://www.eupvsec-proceedings.com/proceedings?paper=7832.

23. Calculated for a mean for 6.4VkW PV arrays.

Appliance and Fixture Replacements

Even when global warming is not the target, policy incentives may unintentionally induce the emission of embodied carbon. Low-flush toilet replacement programs provide an example. Toilets manufactured in the United States before 1980 required five or more gallons of water usage with every flush, an enormous waste of water, literally down the drain. In the mid-1980s, 3.5 gallon models entered the market reducing the waste, only to be bettered by 1.6 gallon models halving potential water usage by 1992. Water conservation became a catalyst for policies regulating "new" construction. Pursuant to the U.S. Energy Policy Act of 1992, 1.6 gallons became the maximum flush volume allowable for non-commercial gravity toilets manufactured and installed after Jan. 1, 1994, thus mandating a major reduction in wastewater across the nation. All new residential installations in the United States as well as aged-out replacements were to be limited to 1.6 gallons of water per flush. Although carbon emissions had nothing to do with this policy, nor were they increased by this policy in any way, that would soon change with the addition of local policies throughout the country. Drought-stricken water districts experienced "light bulb moments," recognizing the obvious advantage of *replacing* older 5- and 3.5-gallon toilets, saving 2 to 3.5 gallons per flush from the millions upon millions of old toilets still in use. Local policy in the form of cash rebates incentivizing toilet swaps made sense. By 2006, Los Angeles alone, through its ultra-low flush toilet initiative had replaced 1.27 million toilets. Similar programs around the country followed suit: more than 72,000 were replaced in Dallas; 135,000 in Atlanta; and a program to replace up to 800,000 in New York City to name a few—saving more than 10 million gallons of water every day.

Unfortunately, there were unintended consequences. These incentivized toilets were not used for new construction projects or broken toilet replacements; they primarily substituted for functioning models that did not require replacement. And similar to policies encouraging small-scale solar installations and appliance replacements, policymakers failed to adequately weigh the upside and downside of promoting toilet replacements. They too stimulate manufacturing and therefore the emission of carbon, spewing a plume of embodied emissions. Manufacturing the porcelain alone for the incentivized replacements in just New York City and Los Angeles emitted more than 100,000 metric tons of CO_2,[24] and this was just a fraction of the emissions

---

24. Calculated using 49.4 $kgCO_2$ embodied emissions per toilet per low-flush bowl and tank toilet fabricated with 30.7kg of vitreous china porcelain; LOS ANGELES DEPARTMENT OF WATER AND POWER, URBAN WATER MANAGEMENT PLAN, Ch. 3 (2020), *available at* https://www.ladwp.com/UWMP..

from similar programs in cities around the United States. The policy deliberations had accounted for none of these emissions.

To be fair, the role of embodied carbon did not gain recognition until the late 1990s, taking a decade to migrate through the building design and construction industry. But having said that, toilet swap programs like these have continued nationwide throughout the last decade, and some into 2018. After 1.27 million toilets had already been replaced in California, and every new residence since 1994 was fitted with 1.6-gallon models or better, California allocated another $6 million in 2015, providing rebates of up to $100 per toilet to encourage replacement of another 60,000 toilets. There are other ways to reduce the water used with every flush. The practice of adding a few old bricks to the tank has been around for a long time. Using a water-filled plastic bottle or plastic bag is even better. For the sake of containing unnecessary carbon emissions in the short term, perhaps new toilet-swap programs can wait—especially as high-efficiency toilets are already mandated for all new construction.

We use a mixed bag of incentives and legislation to achieve environmental goals. Often their metrics lack adequate preconditions, verification or continued monitoring. Few, if any, factor the impact of embodied emissions. Perhaps quantifying embodied carbon piecemeal is too difficult a task, and confronting it contemporaneously even more so. Defaulting to status-quo thinking is far simpler, relying on future abatement by widescale decarbonization. Yet now is the time to make progress, inaction only magnifies the problem.

Where Do We Begin?

Not all policy goals reflect the totality of their impact. Mandating the blending of biofuels with gasoline and diesel fuel, dictating energy-efficient building codes, and allowing millions of acres to burn in the Amazon and in Indonesia all result from specific goals. Whether to reduce emissions or energy consumption, support demand for an agriculture product, or clear land for farming, they all produce unintended consequences as well. Depending on one's perspective—economic, environmental, human welfare, or otherwise—their outcomes may be for better or worse; sometimes with grave long-term ramifications. Thwarting global warming requires capping carbon emissions, yet policies to reduce those emissions may be counterproductive if they neglect the carbon footprint invested to make them operable. During this critical period of climate engagement, the matter-of-fact accumulation of embodied carbon can no longer be taken for granted. Policies and legislation that address emissions must address the emissions footprint they will create. This is where we begin. Including the *environmental cost of embodied carbon* in all material-related decisions is the first step toward reducing net emissions over the next 10 years. This means employing a low-carbon solution, the first line of defense in our battle to retard carbon gap growth on a speedier timeline. This holds true whether designing a new construction project, retrofitting an existing building, or promoting an appliance or product. Whenever feasible, the most effective means to accomplish this goal is simply to extend the useful life of the buildings and appliances that already exist; minimizing the need to create new emissions from manufacturing and construction that trace all the way back to mining and materials processing. Extending useful life through refurbishment, replacement, reconfiguration, or minor expansion not only reduces new embodied carbon expenditures significantly, it can provide substantial financial savings when compared to new construction. Abstaining from carbon-intensive solutions for all of the above includes recycling materials for reprocessing, and designing materials for their eventual reuse.

Recycling, Reuse, Retrofits, and Refurbishment

Beyond designing with wood and low-carbon materials wherever possible, and using as little material mass as feasible, the most effective way to minimize embodied emissions lies in choosing materials that have high recycled content, or that can be reused or refurbished. This is foundational to the concept of a circular economy, where we actively organize a closed-loop system that restores and re-circulates capital, where the expenditure of energy and resources are minimized whether natural, manufactured, social, or otherwise. In the physical world, circularity thrives on managing the flow of materials and energy by extending useful lifetimes, optimizing material use, and minimizing waste. Circularity is more than recycling whatever is suitable, it requires *designing* things to be reusable, to be refurbishable, to be standardized, and to be disassembled easily for reuse or recycling. All of this comes with an environmental bonus, a large reduction in the creation of embodied carbon. Some benefits occur immediately, others longer in term. Recycled content can turn a comparison upside down, and can turn a high-carbon product into a low one, which is especially impactful for aluminum and steel. Manufacturing materials like steel, aluminum, glass, and plastics with a high "recycled" content can provide significant energy savings immediately. The energy expended to manufacture glass can be reduced by nearly 20%;

aluminum by as high as 95%.[25] Of course reusing a component or an entire assembly is even better, providing savings as high as 98%. Saving manufacturing energy eliminates carbon emissions immediately. The benefits from components *designed for reuse* will contribute at the end of a future construction's useful life over the ensuing decades. The future savings will be substantial. *Material circularity* has the potential to cut CO_2 emissions derived from building materials in half by 2050.[26]

Retrofitting and refurbishment clearly outshine demolition and replacement in regard to embodied emissions. Not only do they eliminate the massive carbon embodied in would-be replacement, but the emissions attributable to disassembly and processing for disposal or reuse—there is no comparison. A material economics study published in 2018 emphasizes that "[i]ncreasing the building lifetime is one of the single most effective actions that can be taken to reduce the need for construction materials," noting about 85% of their CO_2 footprint is associated with structural elements, most of which are perfectly sound when buildings are demolished.[27] Extending the life of a structurally sound building with a renovation that *reduces operating energy consumption* as well is doubly effective. The benefits are two-for-one, major savings in both embodied and operating emissions for the price of reducing the latter. By prolonging useful life through refurbishment, retrofit, or alternative use, these benefits derive not only from the materials and components, but from the buildings themselves. It is one of the best ways to eliminate the accretion of embodied emissions from new construction.

When renovations and retrofits are performed solely to acquire future energy savings, where building replacement is not a consideration, emissions embodied in the upgrade play the biggest role, they determine whether or not there is a net benefit, and whether it will be in the short or long term or both. Future gains must be weighed against the embodied emissions incurred to facilitate the upgrade. Are the immediate carbon emission justified for the operating savings anticipated *within* a decade? If not, the retrofit or refurbishment should be re-engineered for a net gain, or delayed to a date when gains in decarbonization make it feasible. Without weighing the value of near-term operating reductions against the embodied emissions invested, that is unlikely to happen. A stunning example of a well-executed plan to achieve energy savings *while* minimizing the embodied emissions invested is the Empire State Building renovation in New York City, completed in 2019.

25. Geoff Milne, Australia's Guide to Environmentally Sustainable Homes (4th ed. 2010).
26. Material Economics, The Circular Economy, A Powerful Force for Climate Mitigation 156 (2018).
27. *Id.* at 152.

Rather than replacing 6,500 heat-leaking double-hung windows, they were removed and rebuilt onsite, reusing 96% of the existing 26,000 glass panes and frames. But there was more. By suspending a low emissivity (low-E) film insert between the existing double panes to reflect thermal radiation, using new spacers, and filling the gap with a mix of krypton and argon gases, the windows were rebuilt as super-insulating units. Their insulating R-value—resistance to heat flow—was increased from R-2 to R-6, cutting the heat gain by more than half.[28]In addition, insulated heat-reflective barriers were installed behind 6,000 perimeter wall radiators. These two retrofits alone were projected to save 8,400 metric tons of carbon emissions over 15 years,[29] yet they added minimal embodied carbon of their own—a substantial energy-saving retrofit with little expenditure of retrofit carbon emissions. Compared to the embodied carbon intensity of new aluminum-framed triple-pane replacements, this effort was an exemplary retrofit with an extremely high net gain. Six additional upgrades and retrofits were made to the lighting, air handling, refrigeration, temperature, and electrical service monitoring. Each was evaluated against its own carbon footprint. The entire program was devised to achieve a 38% energy reduction that would save $4.4 million per year. More importantly, from the standpoint of global warming, it was devised to save 105,000 metric tons of carbon emissions over 15 years while minimizing the carbon emissions invested upfront.

Whether retrofitting an existing building or designing a new one, reducing emissions *by design* requires both an intention and a detailed plan. It entails the materials choice, structural methodology, and the layout of the floor plan and facade: materials, structure, and the environmental interface. Whatever the construction methodology and materials chosen, the physical layout both inside and out will significantly impact the internal conditioning requirements and the energy-related emissions they will create. Much of this relates to the diurnal and seasonal solar cycles, and the predominant wind directions; not just for the overall building envelope, but also for each facet of its facade. The ability to benefit from solar radiation or block it, the ability to benefit from natural light, the ability to benefit from natural ventilation, the ability to benefit from mass are all important things to consider. Without including these factors in the design process, it is likely that solar radiation, the lack of natural light, the lack of natural ventilation, and an inattention to the earth's mass will cause unintended energy-related emis-

28.   Project: Empire State Building, http://www.esbnyc.com/esb-sustainability/project.
29.   Calculated from https://www.esbnyc.com/sites/default/files/ESBOverviewDeck.pdf: "Empire State Building Case Study Cost-Effective Greenhouse Gas Reductions via Whole-Building Retrofits: Process, Outcomes, and What Is Needed Next."

sions. None of this is new; this is standard architecture school curriculum. Unfortunately, in practice, these factors are typically disregarded for expediency, or to achieve a particular aesthetic. Even expressive elements can fulfill performative and structural roles by agency of their surface, substance, and arrangement. Reflecting or projecting heat or light, shadow or color, they interact with environmental phenomena. Such interactions derive from the characteristics of the material properties and assembly, not from the creative expression. *Regardless of a component's primary purpose, it can afford other benefits if designed to do so.*[30]

The ultimate performance of our built environment is a legacy of its design configuration. Environmentally sound design relies on the creativity of enlightened architects, planners and engineers, and a client's goodwill. There exists no shortcut to true sustainable design; it is achieved by adhering to sound environmental principles. But there are formulae to calculate embodied carbon emissions, to weigh the carbon emissions embodied in building design or refurbishment against their projected savings in the near term, and to reduce the hazardous impact of embodied carbon in a timely manner. Environmentally sound design does not necessitate cost increases other than the time required to work through an educated carbon-conscious and energy-efficient design methodology; not just in the initial design schematics, but through final detailing and procurement. This process, however, does need an assist to facilitate action, and a few more tools would help. *Carbon labeling is one prerequisite.*

Informed design and policy decisions rely on Carbon Footprint transparency and reasonably accessible means to compare carbon content alternatives. These are a must. Low-carbon decision-making requires public disclosure of a material or component's carbon content in intelligible and usable terms. It also requires the tools to utilize this knowledge to minimize built-in emissions. Whatever the mechanism to encourage or mandate embodied carbon reductions, it cannot be achieved in a practical manner without the aid of "CO_2-footprint" product labeling (CO_2 labeling), an idea suggested in a report by the Energy, Environment, and Resources Department of The Royal Institute of International Affairs.[31] That report, focused on the potential for low carbon cement and concrete, expressed the need to identify the metrics and measuring methodology to address embodied carbon. At the moment, no coherent means exist to discern the embodied carbon content of the myr-

---

30. Paraphrase from Bill Caplan, Buildings Are for People: Human Ecological Design (Libri Publishing Ltd. 2016).
31. Johanna Lehne & Felix Preston, *Making Concrete Change, Innovation in Low-Carbon Cement and Concrete* (Chatham House Report) (Royal Inst. of Int'l Affairs June 2018).

iad of materials and products that will constitute a built work. Although mass in kilograms or pounds is the primary metric used to indicate a material's carbon content ($kgCO_2$/kg-of-material or $lbCO_2$/lb-of-material) it is an awkward means for comparison, especially to evaluate candidate components and assemblies during a building's design and configuration. As the dimensions and composition of each manufacturer's products vary greatly, computing each material's mass is not an easy task, let alone their composite in a component. Manufacturers' Environmental Product Declarations attempt to shed some light on this problem, but the information they provide rarely helps. Architects, designers, engineers, policymakers, and the public need to know a product's carbon footprint. Mandatory carbon footprint labeling and applets to compile the information are the tools we urgently need.

"CO_2-Footprint" Labeling

In order to facilitate action, all manufacturers of materials, components and assemblies should be required to provide CO_2 labeling that declares their embodied-CO_2eq emissions. We have labeling codes for insulation, labeling codes for fire rating, and labeling codes for energy efficiency, why not for embodied carbon. The labels should be expressed in units that are commonly used by designers, engineers, and product assemblers in their normal practice, perhaps standardized within specific "CO_2 emissions" range groups designated by an alphanumeric code. All primary materials, appliances and building materials should be labeled with their CO_2eq emissions per unit of use. This applies to solar panels, water saving toilets, triple pane windows and the like to evaluate their impact. Though such a measure would specifically target the problem of embodied carbon in the construction sectors, if CO_2 labeling were required for consumer products as well, from toys to home furnishings to automobiles, it would enable an extraordinary reduction in carbon emissions through the power of consumers' choice.

Mandatory labeling of building and construction materials should be made effective by 2025 with a publically available tabulation by product category without cost. Manufacturers should be required to provide labeling values on all product specification sheets, and on promotional material as appropriate. Until then, an effort should be made to combine and standardize whatever data is currently available in a format friendly to architects, engineers, and designers, usable without the need for statisticians or laborious calculations. This is far from rocket science. All buildings require a structure, so labeling for concrete, steel, and wood products is a good place to start; followed by products for cladding, roofing, interior walls, and floors, because aluminum, brick, ceramic, and plastics are high on the emissions list as well. Although this information is available from a variety of organizations, their formats are generally inconsistent and usability is severely lacking.

Web-Based Applets to Compile CO_2 Footprints

The carbon content of a new construction's schematic is a giant puzzle which must be solved on the drawing board, not lamented when built. The carbon intensity of its "bill of materials" must be established in advance to facilitate informed decisions, to enable designs within a carbon cap. The carbon footprint of the materials and assemblies are the puzzle's pieces and CO_2 labeling will help, but we lack a straightforward way to assemble the entire picture. User-friendly online applets and downloadable computer applications would allow footprint data to be gainfully used and "transparency" to become a working tool, ideally without cost. Organizations and agencies such as the, U.S. Green Building Council, NREL, and the U.K. Building Research Establishment (BRE) should create a working group that includes those currently offering programs of this nature. There are several such computation programs already in use with limited capability, their sponsors' knowledge can significantly streamline the development timeline. The NREL already provides programs that estimate the output from solar panel installations free of charge. Why not provide comparable user-friendly programs to estimate embodied carbon?

Without the transparency that CO_2 labeling provides and a meaningful method for user-friendly compilation to enable design strategies based on low-carbon selection, harnessing the building sector's 40% share of emissions by 2030 to 2035 is wishful thinking. The primary data already exists in one form or another, but it needs reformulating for intelligible and widescale use. Computation programs and applications already exist though awkward and laborious to use. The wheel does not need invention, nor does the engine, only to be resized and fitted with an automatic transmission. This too is not rocket science, it is a matter of cooperation, organization, and some detailed hard work. Intelligible CO_2 labeling is of primary importance not only to evaluate and create construction related policies, but to aid other environmental policy decisions as well. Unless the CO_2 footprint of all things considered is readily evident, design decisions, policies and legislation will continue to foster the early release of carbon—unintentionally. And when it comes to the excessive use of concrete and cement, *moral suasion* is another tool that can help.

Motivating Public Opinion

Concrete, the world's most-used construction material is the primary vehicle for the material most responsible for carbon emissions—cement. According to the 2018 Chatham House Report *Making Concrete Change*: "Every year more than 10 billion tonnes of concrete are used."[32] This translates to more than one metric ton of concrete every year for every human alive. Concrete is used to fabricate a large array of products from ornamental flower pots, garden sculptures, and furniture to building ornamentation, roof tiles, facade panels, and institutional interiors. Concrete is an architecture style in itself, which is much admired by corporations, institutions, and developers of upscale residential buildings—even for buildings touted as being "sustainably designed." Buildings are our greatest human canvas. In a given locale they are part of everyone's experience: from afar, up close, and from within. We live in them, work in them, pleasure in them, pray in them, learn in them, and heal in them. They constitute the largest art form other than landscape and infrastructure such as bridges and dams. The temptation to use their height, breadth, volume, and surface as sculpture and ornamentation is inescapable to an architect. When concrete is the material of choice, the price is a high level of carbon emissions.

Sadly, of the two billion metric tons of CO_2 emitted in 2018 by producing cement, only the heat-related emissions, which is *less than half* of the total, might be reduced in future production through fuel source decarbonization. The one billion metric tons of CO_2 produced by chemical changes during the calcination of limestone would still be emitted.[33] More sadly, concrete is still very much in vogue. We can affect a major reduction in immediate carbon emissions by discouraging the unnecessary use of concrete for aesthetic purposes. A few of the many recent examples of such use are noteworthy: the University of Miami's new LEED®-certified Architecture Design Studio with its overhanging concrete roof; Adjaye Associates' *130 William Street*, a 66-story residential tower in New York City clad bottom to top with hand-cast concrete panels; and Zaha Hadid Architects' *One Thousand Museum* residential high-rise in Miami—clad with 4,800 pre-cast concrete panels

---

32. *Id.*
33. IEA, Technology Roadmap: Low-Carbon Transition in the Cement Industry 5 (2018): Cement production involves the decomposition of limestone (calcium carbonate), which represents about two-thirds of the total CO_2 emissions generated in the process, with the remainder of CO_2 emissions being due to combustion of fuels. Thus despite considerable progress on energy efficiency, the use of alternative fuels and clinker replacements, the sector has the second-largest share of total direct industrial carbon dioxide (CO_2) emissions, at 27% (2.2 gigatonnes of carbon dioxide per year [$GtCO_2$/yr]) in 2014. *Id.* at 20 (approximately 1.5 Process + 0.7 Energy = 2.2 $GtCO_2$/yr.).

weighing up to a ton or more each, shipped all the way from Dubai. The complete global list is very long. Concrete as a style, as ornamentation, or as an inexpensive means to fabricate garden furniture and the like promotes the excessive release of carbon emissions during this crucial time. Increasing public awareness of concrete's climate-warming impact, its "un-sustainability," will help dethrone the vogue and its corporate and institutional desirability. Public awareness can discourage the purchase of concrete consumer products as well. The public needs to know, and when it does the demand should fall—selection governs supply-chain replenishment. The "bottom-up" impact of material selection can have a swift and effective impact when promulgated by building codes, policy, and legislation. Moral suasion can provide an additional punch; it is a tool that should be used widely.

If one doubts that using less construction material substantially reduces GHG emissions, the 2008 economic downturn provides an example. So will the tragic impact of COVID-19 when statistics from 2020 and 2021 become available. In the 2008 case, new construction started to unravel from the 2007 downturn on heels of the 2006 housing bubble. Concordantly, cumulative energy-related CO_2 emissions from the United States alone decreased by 700 million metric tons for 2008 and 2009 relative to the average emissions from 2006 and 2007. The slowdown in U.S. cement and steel production for the construction sector[34] was responsible for 32 million metric tons of that reduction, kicking off a real-time supply-chain response.[35] Comparing that with the potential of operating emissions savings, if every U.S. building built in 2008 and 2009 had been designed to use 10% less energy, it would have taken nearly 8 years of operating reductions just to equal the embodied emissions saved from the curtailed production of steel and cement for new construction. Moreover, steel and cement represent only part of the picture; there were additional reductions attributable to site preparation, elements of the facade, the interior, and materials transportation, and construction—all requiring additional years of operating savings to nullify.[36] Rather than rely on economic downturns and happenstance, we can use policy to influence action and shape legislation to determine the timeline. The use of incentives and mandates can effectively reshape cities and suburbs within a single

---

34. Includes the lime they consumed.
35. Reduction of CO_2 emissions from U.S. cement and steel production in 2008 and 2009 from the average production of 2006 and 2007. Computed from U.S. Geological Survey, *Mineral Commodity Summaries* (2007, 2008, 2009, and 2011); EIA, *Emissions of Greenhouse Gas Emissions in the United States* (2009); and U.S. EPA, *Inventory of U.S. GHG Emissions and Sinks: 1990-2010* (1990-2017).
36. Computed from EIA, *Monthly Energy Review* (Mar. 2019); U.S. Geological Survey, *Mineral Commodity Summaries, id.* (2007, 2008, 2009, 2011, and 2019); EIA, *Emissions of Greenhouse Gas Emissions in the United States* (2009); and EIA, *Commercial Buildings Energy Consumption Survey* (2012).

decade, and protect our natural environment. Nevertheless, to thwart global warming they must *account for* their embodied emissions.

Accounting for embodied emissions means weighing the good against the bad, emissions saved within a specific time frame against those emitted before startup—focusing on net carbon emissions over the next 10 years without being assuaged by the future *potential* of decarbonization. *All* such actions should yield a net reduction within their first decade of use. While ignoring embodied carbon may be a necessary evil to achieve transformative change from forward thinking infrastructure—expanding the power grid to incorporate wind and solar farms; battery farms for solar power storage; scrubbing carbon from the atmosphere—postponing or modifying policies with less-extensive benefits, or achieving their goals through other means would be more propitious. The time to act on that reality is now. Knowing what we are up against, heeding enlightened design, requiring carbon transparency, and applying moral suasion—these are steps that will launch a more sustainable future. Much of this is commonsense. Nonetheless, there lies a long road ahead, one that requires us to steward low-carbon action.

Chapter 8

Taking the Roads Less Traveled

Referring to Robert Frost's poem *The Road Not Taken*, Rachel Carson warned that "[t]he road we have long been traveling is deceptively easy, a smooth superhighway on which we progress with great speed, but at the end lies disaster." The one "less traveled by [offers] our only chance to reach a destination that assures the preservation of the earth. The choice is ours to make."[1] We are still traveling Carson's superhighway, the straightest path from ideation to execution, the shortcut to an environmental disaster. The road is lined with promotions for simple solutions, gold stars for feeble certifications and accolades from an admiring press. This journey cannot be reversed, only modified by rework that accumulates an ever increasing footprint. Though late in the day, we stand at yet another fork in the road, a juncture with a clear view of reality, another opportunity to take a path less traveled, a path with smaller footprints, cleaner footprints leading to a brighter future. We still have a choice. The road to environmental design, to energy conservation and to the sustainability of our built environment was laid half a century ago. The methodologies have existed far longer. We have been talking the talk for more than 20 years now —it is time to walk the walk. It is time to follow this less traveled path with speed and a sense of urgency. To do so in a timely manner, we need policy, legislation and an increase in public awareness to kick-start the action. Thus far, voluntary progress has been moribund.

2021

More focused on "carbon" emissions than ever before, the architecture, engineering, and construction (AEC) industries continue to seek "green" building solutions, yet with little evidence of their adoption on a relevant scale. Some AEC professionals cite a lack of client demand,[2] others emphasize the

---

1. RACHEL CARSON, SILENT SPRING 277 (1962).
2. Russell Gold, *"Green" Buildings Struggle to Catch On*, WALL ST. J., Dec. 26-27, 2020 [hereinafter Gold].

industry's accomplishments, exemplified by particular buildings designed with energy-efficient features or structured from wood—singular examples of informed architecture. Despite an industry-wide awakening over the last 10 years, the focus remains on energy-efficient design and solar panels to lower carbon emissions. The need to address "embodied" carbon has received more attention recently, but not traction. Tools to evaluate a building's embodied carbon have become available in the form of databases, computational software, environmental product declarations (EPD) and forums to disseminate current information, though few if any are user-friendly. Such progress is rightfully acknowledged and one may be thankful for the many examples of sustainably designed architecture. Nevertheless, they are few and far between on the global scale. Construction statistics belie any suggestion that carbon footprint conservation is the norm in the building and construction sector, or that the built environment's emissions will be reduced fast enough to achieve the 2030 goals.

To this date, the design of new buildings and renovations demonstrate little indication that embodied-emissions minimization has been "spec'd in"— neither in the United States nor worldwide. Observing new construction in one's urban or suburban surroundings is likely to confirm just more of the same, hallmarked by their designs and the construction materials employed. As the calendar progressed through 2020 to 2021, the headlines were no more encouraging than the year before. They continued to express alarm at the challenge ahead.

"Green" Buildings Struggle to Catch On
Russell Gold, Wall Street Journal, Dec. 26-27, 2020

The American Institute of Architects has for years challenged its members to design buildings to combat climate change, setting a goal to hit "net zero" edifices by 2030. The architects have a long way to go. Last year, 27 of the 19,000 building-design firms owned by AIA members reported meeting their annual mark.[3]

Gold's article referred to *2030 By the Numbers*, the 2019 Summary of the *AIA 2030 Commitment* released in September 2020. The American Institute of Architects (AIA) 2030 Commitment program was established in 2009 to encourage architects, engineers, and design professionals to take "robust

---

3. *Id.*

action" to achieve a carbon-neutral built environment by 2030.[4] Such commitment is voluntary.

The AIA and similar organizations worldwide should be commended for trying. But seven years after the AIA Board of Directors declared that "sustainable design practices" were a "mainstream design intention" in the architectural community, only 682 out of 19,000 AIA member-owned design firms were signatories to the 2030 Commitment. Of those, only 311 firms reported their "predicted" energy use intensity reductions, merely 27 of which met their annual target for the Commitment. With a caveat that "with the climate crisis escalating, more needs to be done,"[5] these results were oddly characterized as a "significant" improvement. Notwithstanding a statistical increase over the two prior years and the best average reduction in the Commitment's history, over 94% of AIA member-owned firms remain aloof. Gold quotes architects who attribute the lack of progress to a variety of factors such as the clients, the contractors, the cost, and the procedural process of building design. Architects "learned a lesson," a "vast majority of clients aren't asking for green structures" and "few contractors have experience on high-performance buildings." Or contractors that do so might add 2 to 3% to the cost of construction. And while many firms meeting the climate challenge "integrate engineering and architecture at the start of the design process," typically, architects design a building *before* they bring on the engineering consultants.[6] Much of this resembles "passing the buck." And though the 2030 goal is to reduce "net" carbon emissions, the initial carbon footprints from "embodied emissions" are not a metric of the 2030 Commitment, even as their importance was highlighted in the 2019 Summary.

> Operational carbon is only one piece of the climate action puzzle for the built environment. In order to meet international targets, the design community will need to embrace embodied carbon in their designs and decision-making.

> Enabling users to track embodied carbon will not impact calculations toward the 2030 fossil fuel and energy reduction targets, but it will allow architects to evaluate the environmental impacts of their designs more accurately.[7]

Notably, Gold quotes Christoph Reinhart, director of the Building Technology Program at the Massachusetts Institute of Technology School of Architecture and Planning: "Architects know how to build very efficient

---

4.   AIA, *2030 by the Numbers: The 2019 Summary of the AIA 2030 Commitment* (Sept. 2020) [hereinafter *2030 by the Numbers*].
5.   *Id.*
6.   Gold, *supra* note 2.
7.   *2030 by the Numbers, supra* note 4.

buildings." This is correct. Architects, however, know how to reduce embodied carbon as well. Numerous paths are available to architects and designers to reduce inherent carbon emissions without requiring a specific client mandate or request for a "green" building. They run the gamut of design methodologies, material palettes, and the invaluable benefit of engineering input during design conception as mentioned above. For the short term, perhaps the next three to five years, choices may be somewhat limited by the construction products currently available. Nonetheless, they offer a broad range of carbon-intensity values versus energy efficiency, and provide an array of containment opportunities for those so inclined.

The granular work to curtail embodied and operating emissions from new buildings and renovations by 2030 is not the only problem, and the remainder of the 2020s faces another hurdle. Containing new *infrastructure* emissions over the next 10 years must be part of the solution.

A Lot of Work Ahead If the U.S. Is to Reach "Net Zero" on Pollution
Brad Plumer, New York Times (National), Dec. 16, 2020

If the United States wants to get serious about tackling climate change, the country will need to build a staggering amount of new infrastructure in just the next 10 years, laying down steel and concrete at a pace barely contemplated today.

Plumer's article addressed the interim report *Net-Zero America: Potential Pathways, Infrastructure, and Impacts*, a newly released two-year Princeton University study. The report reveals the scope of the challenge ahead to attain net-zero emissions by 2050, a goal endorsed by the Biden Administration. The study presented an array of actions essential to achieve 2050 net-zero targets in the United States, including a set of "robust measures" needed *within this decade*. These measures not only express the enormity of the effort required to reach net-zero emissions nationally, but also indicate the challenges for nations worldwide. Only the particulars and timelines differ.

It has been clear for two decades or more that, for the industrialized countries to do something approaching a responsible share of a global effort to limit the average surface temperature increase to 2.0°C, they would need to reduce their emissions of heat-trapping gases by 80 to 100 percent by around 2050. Each year that has passed without countries taking steps of the magnitude needed to move expeditiously onto a trajectory capable of achieving such a goal has increased the challenge that still lies ahead.[8]

The "robust measures" require not only decarbonizing of our energy supply, but also the capture and sequestration of carbon emissions—all long-term correctives. And consistent with other studies, a *Net-Zero America* relies heavily on expanding the supply and use of carbon-free power to electrify transportation, interior heating, and industrial processing. The Princeton report concludes that

> [b]uilding a net-zero America will require immediate, large-scale mobilization of capital, policy and societal commitment, including at least $2.5 trillion in additional capital investment in energy supply, industry, buildings, and vehicles over the next decade [the 2020s] relative to business as usual. Each transition pathway features historically unprecedented rates of deployment of multiple technologies.

While the report notes that a dozen states have pledged to be net-zero by 2050, and that more state and corporate pledges are on the way, it reveals that the study was motivated by the *scant details available* on how such pledges will be achieved. Although the study offers alternative pathways for technological and infrastructure investment, it too fails to directly address their consequential embodied carbon. This follows the common practice of aggregating embodied emissions within macro-statistical assumptions based on supply-side industrial energy consumption.[9] In other words, it attributes a percentage of future industrial energy-related emissions to the products manufactured, subjugating such embodied emissions to future remedy as energy decarbonization is achieved. The study ignores more immediate reductions available from design, materials selection, and conscientious use—those reliably available now and during the next 10 years. The task ahead for the remaining 2020s is no less daunting in 2021. If anything, this study exposes the critical need to address the addition of new emissions *within this decade*—lest we continue to fall behind.

Professional trade associations such as the AIA, the Royal Institute of British Architects, the European Architects' Alliance, the Architects Regional Council Asia, the U.S. Green Building Council, the U.K. Green Building

---

8.   Eric Larson et al., *Net-Zero America: Potential Pathways, Infrastructure, and Impacts (Interim Report)* (Princeton Univ. Dec. 15, 2020) [hereinafter *Net-Zero America*].

9.   *Id.* 4.6. Emissions:
> [T]here are physical emissions. These are traditional emissions associated with the combustion of fuels, and they represent the greenhouse gas emissions embodied in a unit of energy. . . . Physical emissions are accounted for on the supply-side in the supply nodes where fuels are *consumed*, which can occur in primary, product, delivery, and conversion nodes.

Council, the U.K. Building Research Establishment, and the like have awakened to the necessity of low-emissions design and planning. Though these organizations wield significant influence with architects, engineers, and decision-making developers, the actions they propose remain mostly voluntary. For multiple reasons the quest for "voluntary" action has not been successful, neither for widescale decarbonization nor for low-carbon design. Simply stated, voluntary recommendations have not been effective enough to retard annual emissions growth. According to the *2018 Global Status Report*, "mandatory and voluntary building energy codes exist in 69 countries worldwide, but nearly two-thirds of countries still do not have mandatory building energy codes that cover the entire buildings sector." A more sobering statement noted that merely 5 of 197 signatories to the United Nations Framework Convention on Climate Change (UNFCCC) mentioned measures to address a building's embodied carbon; "ambitions to reduce embodied carbon in buildings are in the background."[10]

Broad goals like "peaking" annual emissions by a specific date or eliminating so many megatons of carbon emissions provide scant guidance to architects, engineers, and their commissioning clients. Such generalities offer little help without intelligible national and local guidelines to reduce carbon emissions and minimize embodied carbon. Carbon-reduction goals must be relatable to specific industries and economic sectors to be actionable. Absent relatable objectives, implementing one solution while aggravating a problem with another will persist, triple-pane windows and rooftop solar arrays will continue to satisfy "green building" compliance despite the excess use of cement.

It is too late to ignore embodied emissions and too late to rely on lauding minimal accomplishments and voluntary actions. With only nine years remaining in the 2020s, the AIA is not mincing words regarding the dire need for action, or *who* is responsible for containing the built environment's contributions.

> The consequences of climate change are alarming, but they are not inevitable. Globally, buildings account for 39% of total greenhouse gas (GHG) emissions.
>
> The design industry is largely responsible for eliminating that output. The design sector is at an inflection point; Every action we do not take today compounds our challenges tomorrow. . . .

---

10. IEA & UNEP, GLOBAL STATUS REPORT FOR BUILDINGS AND CONSTRUCTION: TOWARDS A ZERO-EMISSION, EFFICIENT, AND RESILIENT BUILDINGS AND CONSTRUCTION SECTOR (2018).

As a profession, the design community has the responsibility to prioritize and support effective actions to exponentially decelerate the production of greenhouse gases contributing to climate change.

With less than a decade left to meet our industry's 2030 deadline, it is time for every company—and every design professional—to act. . . .

The design industry can still meet the targets—if it acts now.[11]

These are strong words, but useless if they fail to stimulate timely actions. With all due regard for the industry's accomplishments it is essential to face a reality—the past 10 years of discussion and good intention must be turned into action *now*. The book's foregoing examples lay bare a broad range of design and systemic issues which, lacking insight and oversight, still await action to resolve.

The Roads Less Traveled: Stewarding Effective Actions Through Activism, Policy, and Legislation

Whether through incentives, mandates, or moral suasion, policies and legislation that address the carbon intensity of building components will speedily reduce carbon emissions throughout the industrial sector by means of market forces. Though the built environment is constructed and equipped one structure at a time, the mining, processing, and manufacturing of its materials and components are generally accomplished in large lots—long prior to their use. Often released to the atmosphere years before they accrue to a particular building, these emissions are many times more catalytic to global warming than can be mitigated by future reductions in a new building's operating emissions. Decreased demand for carbon-intensive materials in the short term will influence the supply chain for high-carbon building materials, retarding their manufacture and resupply. The ensuing reshuffle of economic components is necessary to attain carbon reduction targets within the required time frame. We can no longer afford to pay for a future reduction of operating emissions with *unchecked* emissions from embodied carbon. At best, we can reduce both, but at a minimum we must address embodied carbon first—it determines the baseline. Emissions transparency will not solve the problem alone; to make headway we must follow it by catalyzing action. When carbon footprint information is available in clear and useful terms, and architects, engineers, manufacturers, and their clients have the tools to

---

11. *See 2030 by the Numbers, supra* note 4.

make intelligent decisions, policies, and legislation can effectively mandate the appropriate guidance.

Mandating Embodied Carbon Limits

Until annual emissions levels have peaked and begun to decline, and zero-carbon energy has become the predominant source of power, temporary mandates can and should regulate the emissions embedded in construction—for buildings and for infrastructure. In other words, embodied carbon caps *by floor area* for residential and commercial buildings are essential, scaled appropriately by use-category and size. Infrastructure projects must be regulated with appropriate standards as well. Though this may sound both draconian and undoable, carbon cap mandates by floor area are already legislated in New York City for "operating" emissions, and in policy elsewhere; it is essential to include comparable caps for "embodied" carbon.

Recognizing the urgency to thwart global warming, New York City, the most populous U.S. city, has established some of the country's most progressive emission reduction policies. Even with a 2007[12] goal of a 30% decrease in greenhouse gas (GHG) emissions by 2030 having achieved significant reductions, the city instituted a more stringent target to reduce emissions by 40%. Acknowledging that earlier gains stemmed largely from the reduced use of high-carbon fuels, New York City officials realized that further improvements would require a more comprehensive approach. Eighty percent of the city's reductions resulted from abandoning coal and improving utility operations. Power plants had already switched from coal to natural gas, and heating systems had been converted from heavy to lighter fuel oil or natural gas under the city's Clean Heat Program. Rising to the continued challenge and acknowledging that "these strategies cannot be replicated," future reductions would be more difficult to achieve,[13] New York mandated action at a higher level. In the words of New York City's *Buildings Technical Working Group*:

> There is growing consensus that the current approach of incremental improvements to the Energy Code's prescriptive requirements for specific building systems will not be sufficient to achieve the necessary carbon reductions in the near-term. Instead, a new energy code must consider the entire building as an integrated system by requiring new buildings and substantial renovations to be designed to a whole building energy performance standard.

---

12. One City Built to Last, Transforming New York City Buildings for a Low-Carbon Future (New York City 80 x 50 Buildings Technical Working Group Report, 2016).
13. *Id.*

Implementing these standards as soon as possible will prevent the need for future retrofits. . . .[14]

In 2019, New York City passed Local Law 97 to update its commitment to reduce GHG emissions by 2050. The law mandated a 40% reduction in city government emissions by 2025 relative to 2005, and citywide by 2030.[15] In addition to establishing the Office of Building Energy and Emissions Performance to administer the program, and legislating prescriptive energy conservation measures to ensure the proper maintenance, repair, and insulation of heating and hot water systems, it instituted *caps on annual building operating emissions* for buildings exceeding 25,000 square feet. The caps become effective in 2024. From 2024 through 2029, the maximum allowable carbon emissions per square foot of gross floor area range from 4.3 to 28 kilograms (kg) of carbon emissions as a function of building use, with an average of 10.5 kg for the 10 applicable use categories. From 2030 through 2034, the maximum allowable range will be reduced to 1.7 to 11.9 kg of emissions with an average of 4.6.

On Earth Day 2019, Mayor Bill de Blasio announced New York City's Green New Deal, declaring that "we are going to introduce legislation to ban the glass and steel skyscrapers that have contributed so much to global warming," meaning those failing to meet the energy-efficiency standards required to reduce emissions. He was not advocating a ban on glass and steel, but referring to their need for appropriate energy-conserving designs. Yet as bold and admirable New York's program has been, two words have carried little weight—"embodied carbon." If capping "operating" emissions is now legislatively mandated, why not emissions from *embodied carbon*?

With new mandates giving "building owners just five years to cut their pollution or face major fines"[16] and requiring solar panels or a green roof on every suitable building, New York clearly recognizes the urgency of acting within this decade. Because embodied carbon is emitted first, it cannot be ignored. It too should be capped by floor area. With nationwide construction of privately owned single-family homes exceeding 900,000[17] annually in 2020 with a median floor area one-tenth New York's 25,000 square foot threshold for capping operating emissions, and the average floor area for

---

14. *Id.*
15. Increasing minimum reductions in citywide emissions from 30% to 40% by 2030 and for city government by fiscal year 2025.
16. Mayor de Blasio Announces New York City's Green New Deal, comment by Director Daniel Zarrilli, OneNYC (Apr. 22, 2019).
17. Press Release, U.S. Census Bureau, Housing & Urban Dev., Monthly New Residential Construction, December 2020 (Jan. 21, 2020) (No. CB21-11).

commercial construction through 2018 below that cap as well,[18] embodied carbon policies for new construction are needed both nationwide and on a global scale appropriate to local national standards. Parameters suited to infrastructure and institutions should be established as well. Putting in place embodied carbon caps initially for a 10-year period, for construction commencing no later than 2025, such policies could be reviewed every 5 years for relevance in conjunction with the level of carbon-free energy used worldwide. As our energy supply becomes decarbonized, so will our annual operating emissions and the embodiment of carbon in future materials. Mandates that cap embodied emissions along with operating emission initiatives would effectively reduce the built environment's carbon gap in the shortest period of time. They need not mandate a particular materials palette, nor a structural methodology. Clients would still have full control to dictate the aesthetics and cost. Countless ways exist to realize a client's program while limiting the project's embodied carbon.

The First Five Years Forward—Working With the Capital We Have

Achieving the ultimate containment of global warming through decarbonization of the global energy supply and carbon sequestration requires a commitment to large-scale mobilizations that include expanding low-carbon and no-carbon energy generation. This also includes energy storage and transmission infrastructures, increasing the use of renewables, and refining the technologies of carbon capture. Progress is happening, but not enough to counter the pace at which human activity continues to increase the atmospheric concentration of carbon dioxide (CO_2). The implementation of policy and legislative intervention can produce stabilizing results within the current decade, but it takes several years to formulate policy and enact legislation before they can begin to gain traction. Until new policies and legislation can be put in place to effect a more immediate change, or more user-friendly tools are available to assist, it is up to the pool of individual actors who possess the ability to employ low-carbon design methods and materials, and to influence their use by others. Those actions are not dependent on large-scale mobilizations, trillion dollar allocations, or long-term policy implementation. They can be accomplished through the impetus of designers, detailers, and constructors, and augmented by public demand. To increase our probability of success, we must set in motion the professionals who design and construct

---

18.   EIA, *2018 Commercial Buildings Energy Consumption Survey (Preliminary Results)* (Nov. 2020).

our built environment, as well as those who drive its design by demand—the developers, homeowners, renters, businesses, and institutions.

What can we achieve within a five-year period, to reset the foundation for this decade's gains, to avoid moving the goalposts once again, to avoid undercutting what could have been significant net gains? How can we reduce new "net" carbon emissions within the current decade, fully accounting for embodied emissions, those routinely taken for granted in the name of obtaining *future* decarbonization? Millions of new buildings and renovations are in various stages of design or detail specification, and millions more will be constructed every year henceforth commencing next year. All of their materials and component parts will have released *new* carbon emissions during processing and manufacturing. They cannot await future energy supply decarbonization, and some of their emissions will have come from intrinsic chemical reactions. Architects, engineers, developers, and commissioning clients are making design decisions as you read this book. The resulting manufacturing emissions may be *released elsewhere*, but cannot be rationalized by a locale's tunnel vision, indifference when not in one's own backyard. New York City's aggressive program for a "low-carbon future" concerns emissions released in *New York*. The tally excludes emissions attributable to building components procured from across the United States or abroad—from steel to glass to solar panels. Unfortunately, global warming knows no borders. Atmospheric carbon is not contained by engendering its release in another locale. Gains within the next 5 years are crucial to the 10-year timeline; stabilizing "net" emissions by 2030 is crucial to the world's 2050 goals. How do we move forward to reduce new emissions in real time over the next five years, before we have the advantages of user-friendly computation tools, CO_2-*footprint* labeling, new policy, and legislation? All of these take time to develop and initiate. We must walk the walk first with the capital we have. *This is the engine for change that is available now.*

To successfully tackle the upfront surge of emissions from each new construction or addition within a five-year time frame, we must activate those currently able to influence design specifications, options, and decisions—our human capital. This pool of decision making designers and specifiers, the architects, designers, and engineers who layout the options is key even though developers and clients ultimately determine the outcome. The sheer number of registered architects alone reflects their potential impact. According to the National Council of Architectural Registration Boards (NCARB), in the United States, more than 116,000 architects were registered at the start of 2020.[19] The Architects' Council of Europe represents the interests of more than 500,000 architects in Europe.[20] The International Union of Architects estimates approximately 1.3 million architects worldwide.[21] Adding designers and engineers to this sizeable pool of architects greatly magnifies the potential impact. Actions of these specifiers alone can substantially reduce the release of new embodied emissions over the next 5- and 10-year timelines, the most immediate results fostered by those currently influencing design decisions. Energy-efficient design is already a mainstream goal for these professionals. What lacks is a sense of *urgency* to minimize embodied carbon footprints, a stimulus for *immediate* action.

Until user-friendly tools become available and policies kick in, the shortest route to progress lies in outreach and activism—activating architects, designers, and engineers to reduce the embodied footprint of their next structure built. This is our only means to initiate a reduction in new embodied carbon from the buildings and construction sectors over the next several years. Given the lack of tools available, immediate headway may seem unrealistic, but only in a relative sense. With the availability of proper tools, opportunities to reduce embodied emissions will increase exponentially. CO_2-footprint labeling, user-friendly product emissions databases, and a user-friendly means to facilitate computation will enhance whole-building analysis in detail, shedding light on the specific embodied emissions impact of a wide range of building materials, components, fixtures, interior finishings, furniture, and equipment. Nevertheless, in the interim, absent these tools, we can still rise to the challenge of "urgency" by making heuristic selections based on industry-wide common knowledge. We can avoid succumbing to a lack of *action* and alter the status quo.

Whether via structural choices of wood versus steel versus concrete, roofing, and cladding with materials other than concrete or ceramic tiles or panels, or minimizing the use of concrete block or not cladding them with brick, we can lessen new embodied emissions in construction projects and renovations. Design professionals understand the common "relative" emission comparisons. The value derived from minimizing the excessive use of concrete or by avoiding its use for artistic embellishment is just common sense. Simply eliminating the excess use of materials by design helps facilitate containment. This is not a wheel that needs reinvention with each new project. Once constructed and reviewed, a project's construction methodology and materials

19. National Council of Architectural Registration Board, *NCARB by the Numbers 2020*, Washington, D.C.
20. Architects' Council of Europe, *2019 Annual Report & 2020 Outlook*.
21. Union Internationale des Architectes, *Our Story* (June 24, 2019).

pallet constitute a pattern for reemployment in an architect's or designer's toolbox, perhaps refined with each new project. That is what professionals do. In other words, the first step is making *conserving* embodied carbon *the norm*, the best practice—rather than a "green" or "sustainable" option. Minimizing designed-in embodied carbon in this manner by 5 to 15% will go far toward suppressing emissions escalation within the 2020 decade. It requires activating the design sector to utilize their existing knowledge and training.

Advocacy is the means to stimulate such action, but it must be pursued on a broad scale. The AIA and other such professional organizations worldwide have tried to engage their members in this pursuit with insufficient large-scale success. News media, trade press, social media, and educators are some of the vehicles that can boost the demand by spotlighting the requirement to reduce upfront emissions now—urging architects, designers, engineers, and developers to select low-carbon pathways and soliciting public demand as well. The public is largely unaware of the importance of its "embodied carbon footprint" and its immediate impact on the global warming timeline. Public perception of "carbon footprint" commonly relates to fossil fuel consumption by vehicles, and to energize light bulbs, heaters, and air conditioners rather than the emissions released in their manufacture. Educating the citizenry regarding embodied carbon and the crucial 10-year emissions timeline is key to reshaping current practices and avoiding false choices. This requires instruction in what constitutes embodied emissions, their origins in all objects manufactured, and their immediate influence on the increase of atmospheric CO_2. The public must be urged to press for more energy-efficient buildings, and more importantly at this moment, to the "urgency" of reducing *new* "embodied" emissions. Popular demand drives a large portion of the global economy's physical production and construction. Advocacy is the mechanism to enlighten and activate those capable of effectuating carbon reductions in real time, and those who will influence the policies and legislation to come.

The AIA and other architecture and construction trade organizations and institutions are uniquely situated to quickly activate their grass-roots elements. The AIA's tentacles extend far beyond their 19,000 member-owned firms to their 90,000+ members in local and regional chapters worldwide. Reachable via email and text, its national organization has the ability to transmit this message of "urgency" directly, and to their local chapters' agendas at their next monthly meetings, a direct call to action to the very people who participate in the specification of embodied carbon. The AIA's bully pulpit can identify and address unnecessary embodied carbon as an environmental poison. Once prioritized, the benefits of conserving embodied carbon can begin to accumulate within a few years, without awaiting new technologies or substantial decarbonization of the global energy supply.

The 10-Year Timeline

The first five years afford the opportunity to stimulate voluntary carbon conservation by educating the influencers, policymakers, and legislators to embodied carbon's accelerating influence on the global warming timeline. This period is essential to developing embodied carbon-conscious policies and generating local legislation, to refining user-friendly tools, and to instituting CO_2-footprint labeling. Achieving the 10-year goals requires all of these to be in place along with an enlightened public. The potential to successfully incentivize or mandate embodied-carbon conservation, and to effectively design for it, can feasibly produce significant emissions conservation by 2030. Evaluating the upfront investment of embodied emissions both geographically and by time frame will foster the most appropriate technological solutions. When no longer taken for granted, accounting for embodied emissions will become part of the solution.

Exclusive of investments in low-carbon energy storage, transmission infrastructure, carbon capture, and sequestration, if a "net" mitigation of operating emissions cannot be achieved within the 2020 decade—*inclusive* of their embodied emissions—such investments will continue to push the atmospheric concentration of CO_2 toward the global tipping point. Emissions resulting from the embodied portion will not fade away. When both operating and embodied emissions are included in respect to their individual timelines, and become the foundation for decision making, the risk of negating their perceived gains in real time can be minimized. As such, we should not put the cart before the horse, but foster significant gains in the short term while we invest in the long-term solutions. We must kick-start the reduction of newly accumulated embodied carbon with the capital we have, buying time for new policies, carbon labeling, and legislation to kick in.

The Choice Is Ours to Make

Research for this book began on Earth Day 2018. This last chapter was originally drafted on Earth Day 2020 while COVID-19 broadened its attack on humanity worldwide. During those two years, recognition of embodied carbon's role in accelerating global warming began to emerge, showcased in architecture, engineering, and construction circles—yet absent action on a viable scale while our window of opportunity continues to close. Nevertheless, by the time this book is published we might see a brief hiatus in the growing concentration of atmospheric carbon. Years later, some may misattribute this slower growth to efforts to reduce carbon emissions, perhaps to energy-saving efficiencies or building design—once again misplaced praise. But it will have resulted from the global impact of COVID-19 spreading across the planet, halting human activity in mere months with the ensuing severe and immediate reductions in transportation, manufacturing, and construction. Atmospheric CO_2 will have plateaued briefly from immediate and asymptotic reductions in the generation of embodied GHGs and operating emissions.

With the cost of more than four million lives and a crushing blow to the global economy, humankind will have experienced firsthand the fragility of life as known on this planet. We will have experienced too how quickly we can react in concert to thwart an existential threat, or alternatively, let it grow more consequential through inaction. Endeavoring to retard global warming without thwarting embodied carbon, is like striving to contain COVID-19 without wearing a mask. The time to parse self-serving messaging has come; the time to use commonsense is upon us. These are the roads less followed before—*now is the time to follow them.*

Millions upon millions of residences will be built as individual dwellings, or in low-rise configurations and towers; offices, manufacturing facilities, and institutions will be built as well. Hopefully, war-torn countries will be rebuilt from scratch in the coming decades. All afford an opportunity to halt the rampant proliferation of embodied carbon, to reduce its heavy footprint to a light profile. Maintain the carbon content of what you possess, sustain its embodiment for a long, long time. Choose the carbon footprint of what you acquire with caution, the only attribute it embodies is prior emissions. Therein lies our future and that of our children; give rise to heat-trapping gases more sparingly, the emissions we release today will never be nullified, even in our great-great-great grandchildren's lifetime. During the next decade, more than 60%[22] of the carbon emissions accrued from additions to the built environment will likely emanate from embodied carbon, *from*

---

22. Based on IEA GLOBAL STATUS REPORT 2020 building and construction sector data for 2019 emissions, prorated for 77 billion square feet of new construction. *See* IEA & UNEP, GLOBAL STATUS REPORT FOR BUILDINGS AND CONSTRUCTION: TOWARDS A ZERO-EMISSIONS, EFFICIENT, AND RESILIENT BUILDINGS AND CONSTRUCTION SECTOR 2020 (2020). *See also* IEA, PERSPECTIVES FOR THE CLEAN ENERGY TRANSITION: THE CRITICAL ROLE BUILDINGS (Apr. 2019), with a building sector emissions rate for a 235 billion square feet of global building stock and IEA & UNEP, GLOBAL STATUS REPORT FOR BUILDINGS AND CONSTRUCTION: TOWARDS A ZERO-EMISSION, EFFICIENT, AND RESILIENT BUILDINGS AND CONSTRUCTION SECTOR 2017 (2017).

buildings, renovations, and infrastructure yet to be built—yet to be designed. With forethought, we can thwart global warming and climate change with each project planned. The time has come to act in earnest before atmospheric carbon passes the tipping point. Perhaps this discussion will help.

About the Author

With an engineer's understanding of sustainability, a passion for people-friendly building design, and a 34-year career in high-technology, Bill Caplan then researched the built environment from a human and environmental perspective for more than a decade, contrasting designers' claims with their ecological veracity. A sober look at the reality of our efforts to contain global warming and the public's self-delusion about "green" and "sustainable" living inspired Bill to write *Thwart Climate Change Now*.

Mr. Caplan's tenure at the multi-national instrumentation company he founded spanned high technology projects from the U.S. space and defense programs to decoding the human genome. Shifting focus to the built environment in 2006, he enrolled in Pratt Institute's Graduate School of Architecture. He subsequently founded ShortList_0 Design Group LLC, seeking to promote an integrated design process to unify sustainable technology and architectural form. In 2016, he published *Buildings Are for People: Human Ecological Design*, a holistic approach to creating user- and community-friendly buildings that are sustainably designed. *Contrast 21c: People & Places* followed in 2018, a photographic essay about people and places in Cambodia, Laos, and Vietnam highlighting the disparities between rural and urban areas, and their struggles to adapt to a 21st-century environment.

Bill Caplan holds a Master of Architecture from the Pratt Institute Graduate School of Architecture and a Materials Engineering degree from Cornell University College of Engineering. He has guest lectured at Cornell University's College of Human Ecology, Cornell's College of Engineering, and to the architecture profession.

Acknowledgments

I would like to acknowledge the many researchers, academics, and scientists who strive to uncover the facts regarding carbon emissions, and make them available without sugar-coating. Thank you. Without you, this book would have been impossible. And thank you to those architects and engineers who endeavor to create an environmentally sound and sustainable world by applying science, by examining cause and effect without succumbing to the indiscriminate labels of sustainable and green. You are essential to thwarting global warming.

I wish to thank John G. Robinson, Ph.D., Joan L. Tweedy Chair in Conservation Strategy at the Wildlife Conservation Society, for his detailed analysis of the book's argument and a review of the science. John, your considerable time spent notating the draft text and in phone conversations helped to add clarity to its final revision. The specificity of your insights helped to enhance the thesis's logic and flow. For all of that I am grateful.

After reading the manuscript, Ken Berlin, President and CEO of The Climate Realty Project, initiated a dialogue that continued for several months, meticulously raising a broad range of stakeholder and timeline perspectives that must be addressed. This included the pressing need for more solar installations to reduce fossil fuel demand, and the reality of weighing less than perfect choices against their potential long-term gains and ancillary community benefits. Ken, thank you for the time taken and the thoroughness of the arguments you expect the book's thesis to encounter. Our discussions enabled me to address them directly in the text. And thank you to The Climate Reality Project for training more than 31,000 Climate Leader Activists who mobilize communities in 170 countries to press for clean energy policies.

A special thank you to Paula Luria Caplan, an urban planning and policy consultant with decades of experience at the New York City Department of City Planning, the Office of Bronx Borough President, and elsewhere. Paula's understanding of local communities, city government, and the path from policy to legislation was indispensible to their expression in this text. Full disclosure, Paula Luria Caplan is my wife, which means she lent an ear to every aspect of this text from concept to completion over the last three years.

Finally, thank you Environmental Law Institute Press for bringing the impact of "embodied carbon" on global warming to public awareness. And to

Rachel Jean-Baptiste, Associate Vice President, Communications and Publications, and to the staff of the Environmental Law Institute and ELI Press, thank you for expediting this book's publication in such a short time frame. Your efforts are much appreciated.

Principle References

Andrew A. Lacis et al., *Atmospheric CO$_2$: Principal Control Knob Governing Earth's Temperature*, SCIENCE, Oct. 15, 2010.

ANNUAL ENERGY OUTLOOK, US ENERGY INFORMATION ADMINISTRATION (2019).

Architect: The Carbon Issue, THE JOURNAL OF THE AMERICAN INSTITUTE OF ARCHITECTS (Jan. 2020).

BILL CAPLAN, BUILDINGS ARE FOR PEOPLE: HUMAN ECOLOGICAL DESIGN (Libri Publishing Ltd. 2016).

Building Design & Construction, White Paper on Sustainability (Nov. 2003).

Christian Breyer & Jürgen Schmid, Population Density and Area Weighted Solar Irradiation: Global Overview on Solar Resource Conditions for Fixed Tilted, 1-Axis and 2-Axes Pv Systems, https://www.eupvsec-proceedings.com/proceedings?paper=7832.

CENTRAL POPULATION AND HOUSING CENSUS STEERING COMMITTEE, THE 2009 VIETNAM POPULATION AND HOUSING CENSUS: MAJOR FINDINGS (Hanoi, June 2010).

COLIN R. GAGG, CEMENT AND CONCRETE AS AN ENGINEERING MATERIAL: AN HISTORIC APPRAISAL AND CASE STUDY ANALYSIS (Elsevier Ltd. 2014).

Circular Ecology Ltd, *Inventory of Carbon and Energy (ICE)*, Database Version 3.0 Beta (Aug. 9 2019 & Nov. 7, 2019).

City of Cincinnati, Ohio, Property Tax Abatement for Green Buildings, Ordinance No. 502-2012 (2013).

Council on Tall Buildings and Urban Habitat, *Height Criteria for Measuring & Defining Tall Buildings*, skyscrapercenter.com and ctbuh.org.

CREATING MARKETS FOR CLIMATE BUSINESS: AN IFC CLIMATE INVESTMENT OPPORTUNITIES REPORT (IFC 2017).

David Feldman & Robert Margolis, *Q42018/Q1 2019 Solar Industry Update* (NREL/PR-6A20-73992) (May 2019).

Ed Dlugokencky & Pieter Tans, NOAA/ESRL, *Earth System Research Laboratories' Global Monitoring Laboratory of the National Oceanic and Atmospheric Administration*, www.esrl.noaa.gov/gmd/ccgg/trends.

Eric Larson et al., Net-Zero America: Potential Pathways, Infrastructure, and Impacts (interim report) (Dec. 15, 2020).

ELENI SOULTI & DAVID LEONARD, THE VALUE OF BREEAM: A REVIEW OF LATEST THINKING IN THE COMMERCIAL BUILDING SECTOR (BRE Global Ltd, 2016).

Empire State Building Case Study, Cost-Effective Greenhouse Gas Reductions via Whole-Building Retrofits: Process, Outcomes, and What Is Needed Next, https://www.esbnyc.com/sites/default/files/ESBOverviewDeck.pdf.

GALEN BAROSE & NAIM DARGHOUGH, TRACKING THE SUN: PRICING AND DESIGN TRENDS FOR DISTRIBUTED PHOTOVOLTAIC SYSTEMS IN THE UNITED STATES (Lawrence Berkeley Nation Laboratory Oct. 2019).

GEOFF MILNE, AUSTRALIA'S GUIDE TO ENVIRONMENTALLY SUSTAINABLE HOMES (4th ed. 2010).

Global Methane Initiative, *Global Methane Emissions and Mitigation Opportunities*, https://www.globalmethane.org/documents/gmi-mitigation-factsheet.pdf.

INTERGOVERNMENTAL PANEL ON CLIMATE CHANGE, 2006 IPCC GUIDELINES FOR NATIONAL GREENHOUSE GAS INVENTORIES (2006).

INTERGOVERNMENTAL PANEL ON CLIMATE CHANGE, CLIMATE CHANGE 2014: MITIGATION OF CLIMATE CHANGE, *Ch. 9, Buildings & Ch. 12, Human Settlements, Infrastructure, and Spatial Planning* (Contribution of Working Group III to the Fifth Assessment Report of the Intergovernmental Panel on Climate Change) (Cambridge University Press 2014).

INTERNATIONAL ENERGY AGENCY, ELECTRICITY INFORMATION OVERVIEW (2020).

INTERNATIONAL ENERGY AGENCY, ENERGY EFFICIENCY INDICATORS: FUNDAMENTALS ON STATISTICS (2014).

INTERNATIONAL ENERGY AGENCY, ENERGY TECHNOLOGY PERSPECTIVES 2017: CATALYSING ENERGY TECHNOLOGY TRANSFORMATIONS (2017).

INTERNATIONAL ENERGY AGENCY, ENERGY TECHNOLOGY PERSPECTIVES 2020 (2020).

International Energy Agency, Global Energy and CO_2 Status Report (2019).

International Energy Agency, Perspectives for the Clean Energy Transition: The Critical Role of Buildings (2019).

International Energy Agency, Renewables Information: Overview (2019).

International Energy Agency, Renewables Information: Overview (2020).

International Energy Agency, Technology Roadmap Low-Carbon Transition in the Cement Industry (2018).

Johanna Lehne & Felix Preston, Energy, Environment and Resources Department, Royal Institute of International Affairs, Making Concrete Change, Innovation in Low-Carbon Cement and Concrete (Chatham House Report, June 2018).

Jos G.J. Olivier et al., Trends in Global CO_2 and Total Greenhouse Gas Emissions: Summary of the 2019 Report (PBL Netherlands Environmental Assessment Agency, The Hague 2019).

Jos G.J. Olivier et al., Trends in Global CO_2 and Total Greenhouse Gas Emissions: Summary of the 2020 Report (PBL Netherlands Environmental Assessment Agency, The Hague 2020).

Kris Maher, *Emissions of Carbon Climb 3.4%*, Wall St. J., Jan. 9, 2019, at A3, *also available at* https://rhg.com/research/preliminary-us-emissions-estimates-for-2018/.

Los Angeles Department of Water & Power, Urban Water Management Plan, Ch. 3 (2020), *available at* https://www.ladwp.com/UWMP.

Material Economics, The Circular Economy: A Powerful Force for Climate Mitigation (2018).

Mayor de Blasio Announces New York City's Green New Deal, comment by Director Daniel Zarrilli, April 22, 2019, https://www1.nyc.gov/office-of-the-mayor/news/211-19/transcript-mayor-de-blasio-new-york-city-s-green-new-deal.

Memorandum of Understanding between the United States Environmental Protection Agency and the American Institute of Architects (Feb. 10, 2005), https://archive.epa.gov/greenbuilding/web/html/aia-mou.html.

NASA Earth Observatory, *The Carbon Cycle*, https://earthobservatory.nasa.gov/features/CarbonCycle/page1.php.

NASA Goddard Institute for Space Studies, Forcings in GISS Climate Mode, Well-Mixed Greenhouse Gases, *Historical Data*, https://data.giss.nasa.gov/modelforce/ghgases/Fig1A.ext.txt.

National Institute of Building Sciences, A Common Definition for Zero Energy Buildings (Sept. 2015).

National Renewable Energy Laboratory, Life Cycle Greenhouse Gas Emissions From Solar Photovoltaics (NREL/ FS-6A20-56487 (Nov. 2012).

National Renewable Energy Laboratory, PVWatts® Calculator, https://pvwatts.nrel.gov/.

New York City, Mayor's Office of Sustainability, New York City's Roadmap to 80 x 50 (Sept. 2016).

New York City, Mayor's Office of Long-Term Planning and Sustainability, One City: Built to Last: Transforming New York City's Buildings for a Low-Carbon Future (Sept. 2014), http://www.nyc.gov/html/builttolast/assets/downloads/pdf/OneCity.pdf.

One City Built to Last, Transforming New York City Buildings for a Low-Carbon Future (New York City 80 x 50 Buildings Technical Working Group Report, 2016).

Press Release, Associated Press, Navigant Research Report, Global Building Stock Database 1Q 2018 (Apr. 26, 2018).

Press Release, International Energy Agency, Global Solar PV Market Set for Spectacular Growth Over Next 5 Years (Oct. 21, 2019).

Press Release, IPCC, Summary for Policymakers of IPCC Special Report on Global Warming of 1.5ºC, Oct. 8, 2018.

Press Release, Wood Mackenzie Power & Renewables and the Solar Energy Industries Association, United States Surpasses 2 Million Solar Installations, May 9, 2019, https://www.seia.org/news/united-states-surpasses-2-million-solar-installations.

Press Release, U.S. Green Building Council, Green Building Accelerates Around the World, Poised for Strong Growth by 2021, Nov. 13, 2018.

Press Release, U.S. Green Building Council, U.S. Green Building Council Announces Annual LEED Homes Awards, Recognizing Residential LEED Projects Elevating the Living Standard Through Sustainable Design, June 20, 2019.

Press Release, U.S. Green Building Council, U.S. Green Building Council Releases Annual Top 10 States for LEED Green Building Per Capita, Feb. 2, 2018.

RACHEL CARSON, SILENT SPRING (Houghton Mifflin Co. 1962).

REN21 SECRETARIAT, PERSPECTIVES ON THE GLOBAL RENEWABLE ENERGY TRANSITION, TAKEAWAYS FROM THE REN21 RENEWABLES 2019 GLOBAL STATUS REPORT (2019) (ISBN 978-3-9818911-7-1).

REN21 SECRETARIAT, RENEWABLES 2019 GLOBAL STATUS REPORT (2019).

REN21 SECRETARIAT, RENEWABLES 2020 GLOBAL STATUS REPORT (2020).

Report of the United Nations Conference on Human Environment, Stockholm, June 5-16, 1972.

UNITED NATIONS, DEPARTMENT OF ECONOMIC AND SOCIAL AFFAIRS, POPULATION DIVISION, WORLD URBANIZATION PROSPECTS: THE 2018 REVISION 1, 9 (ST/ESA/SER.A/420) (2019).

United Nations Conference on Environment & Development, AGENDA 21, Rio de Janeiro, Brazil, June 3-14, 1992.

UNITED NATIONS ENVIRONMENT PROGRAMME, EMISSIONS GAP REPORT 2018 (Nov. 2018).

UNITED NATIONS ENVIRONMENT PROGRAMME, EMISSIONS GAP REPORT 2019 (Nov. 2019).

UNITED NATIONS ENVIRONMENT PROGRAMME, EMISSIONS GAP REPORT 2020 (Nov. 2020).

UNITED NATIONS ENVIRONMENT PROGRAMME, GLOBAL STATUS REPORT 2020: TOWARDS A ZERO-EMISSION, EFFICIENT AND RESILIENT BUILDINGS AND CONSTRUCTION SECTOR (2020).

United Nations Environment Programme, *10 Things to Know About the Emissions Gap 2019*, Nov. 26, 2019, at https://www.unep.org/news-and-stories/story/10-things-know-about-emissions-gap-2019.

United Nations Environment Programme & International Energy Agency, Global Status Report 2017: Towards a Zero-Emission, Efficient and Resilient Buildings and Construction Sector (2017).

United Nations Environment Programme & International Energy Agency, Global Status Report 2018: Towards a Zero-Emission, Efficient and Resilient Buildings and Construction Sector (2018).

United Nations Environment Programme & International Energy Agency, Global Status Report 2019: Towards a Zero-Emission, Efficient and Resilient Buildings and Construction Sector (2019).

United Nations Paris Agreement, Dec. 12, 2015, art. 2.1, §1, 55 I.L.M. 743, *available at* http://unfccc.int/files/ess.

U.S. Census Bureau, Characteristics of New Housing, https://www.census.gov/construction/chars/highlights.html.

U.S. Census Bureau, New Residential Construction, Annual Characteristic of New Housing for 2016-2018, https://www.census.gov/construction/chars/historical_data/.

U.S. Census Bureau & U.S. Department of Housing and Urban Development, Monthly New Residential Construction (Sept. 2019) (Release No. Cb19-158).

U.S. Census Bureau & U.S. Department of Housing and Urban Development, Monthly New Residential Construction (Dec. 2020) (Release No. Cb21-11).

U.S. Census Bureau & U.S. Department of Housing and Urban Development, New Residential Construction, https://www.census.gov/construction/nrc/index.html (Aug. 16, 2019).

U.S. Department of Energy, Office of Energy Efficiency and Renewable Energy, The Greening of the White House, Six Year Report (DOE/EE-0000) (Nov. 1999).

U.S. Energy Information Administration, Commercial Buildings Energy Consumption Surveys for 2000-2012, *available at* https://www.eia.gov/consumption/commercial/.

U.S. Energy Information Administration, Electric Power Annual 2018 (Oct. 2019, Revision Notice, Dec. 9, 2019).

U.S. ENERGY INFORMATION ADMINISTRATION, EMISSIONS OF GREENHOUSE GAS EMISSIONS IN THE UNITED STATES 2009 (March 2011).

U.S. Energy Information Administration, Monthly Energy Reviews (March 2019, Nov. 2019, March 2020 & Jan. 2021).

U.S. ENERGY INFORMATION ADMINISTRATION, U.S. ENERGY-RELATED CARBON DIOXIDE EMISSIONS 2017 (Sept. 2018).

U.S. ENVIROMENTAL PROTECTION AGENCY, GLOBAL ANTHROPOGENIC NON-CO_2 GREENHOUSE GASES: 1990-2030 (Revised Dec. 2012).

U.S. Environmental Protection Agency, *Green Building* (updated Feb. 20, 2016), https://archive.epa.gov/greenbuilding/web/html.

U.S. Environmental Protection Agency, *Greenhouse Gas Equivalencies Calculator*, https://www.epa.gov/energy/greenhouse-gas-equivalencies-calculator.

U.S. ENVIRONMENTAL PROTECTION AGENCY, INVENTORY OF GREENHOUSE GAS EMISSIONS AND SINKS: 1990-2017 (EPA 430-R-19-001) (Apr. 2019).

U.S. Geological Survey, Mineral Commodity Summaries (2007, 2008, 2009, 2011, 2018, and 2019).

VICTOR OLGYAY, DESIGN WITH CLIMATE: BIOCLIMATIC APPROACH TO ARCHITECTURAL REGIONALISM (Princeton University Press 2015).

Vincent J.L. Gan et al., *A Comparative Analysis of Embodied Carbon in High-Rise Buildings Regarding Different Design Parameters,* 161 J. CLEANER PRODUCTION 663-75 (2017).

VITRUVIUS, THE TEN BOOKS OF ARCHITECTURE (Morris Hicky Morgan trans., Dover Publications, Inc. 1960) (unaltered and unabridged republication of the first edition published by the Harvard Press, 1914).

WORLD COMMISSION ON ENVIRONMENT AND DEVELOPMENT, OUR COMMON FUTURE (Oxford University Press 1987).

Index